History of the Earth Sciences
during the Scientific
and Industrial Revolutions
by D. H. Hall

History of the Earth Sciences during the Scientific and Industrial Revolutions

with Special Emphasis on the Physical Geosciences

by D. H. Hall

ELSEVIER SCIENTIFIC PUBLISHING COMPANY
Amsterdam — Oxford — New York 1976

ELSEVIER SCIENTIFIC PUBLISHING COMPANY
335 Jan van Galenstraat
P.O. Box 211, Amsterdam, The Netherlands

AMERICAN ELSEVIER PUBLISHING COMPANY, INC.
52 Vanderbilt Avenue
New York, New York 10017

ISBN: 0-444-41440-1

Printed in The Netherlands

ACKNOWLEDGEMENTS

The author wishes first of all to acknowledge the intellectual ancestry of the approach to science taken in the present book. This ancestry lies in those approaches to the history of science in which the social relations of science are recognized. In particular, the author is indebted to J.G. Crowther and J.D. Bernal, especially to the enlightening discussions of science and society afforded by Mr. Crowther, who also gave valuable advice in the beginning stages of the present book.

Many people have given assistance in the preparation of the book. My wife, Esther, and children, Judy, Bernie and Norman have provided assistance in typing and research is much appreciated; their support (before 1972), free of economic demands, was much needed in a project which had to survive its initial stages in a period of sparse assistance for projects of this type. Research assistants David Bellhouse and Jim Bamburak contributed creatively to the book. Mrs. S.D. Fay typed the manuscript and worked loyally to provide additional service in the research and editorial fields. Mr. R. Pryhitko responded creatively to the needs for high quality figures. From 1972-5, the project was supported by a generous grant from the Canada Council.

The author wishes to acknowledge the following authors and publishers for permission to reproduce figures:

Cambridge University Press for permission to reproduce Figure 5 (our Fig. 1.2) from CLARK, J.G.D., 1969. World prehistory: a new outline (2nd edition). Cambridge University Press, London, 331 pp.

American Geophysical Union for permission to reproduce Figure 8 (our Fig. 2.1) from E.H. Vestine's The Survey of the Geomagnetic Field in Space. March 1960, Transactions, American Geophysical Union, 41(1).

The Royal Society and Dr. E.C. Bullard for permission to reproduce our Figure 2.2 from this paper in the Philosophical Transactions of the Royal Society of London, 243: 79.

The American Geophysical Union for permission to reproduce our Table 2.1 from HELLMANN, G., 1899. The beginnings of magnetic observation. Terrestrial Magnetism and Atmospheric Electricity, IV(2): 73—86.

VI

M.J.S. Innes for permission to reproduce Figure 6 from MILLER and INNES, 1953. Application of gravimeter observations to the determination of the mean density of the earth and of rock densities in mines. Publications of the Dominion Observatory, Ottawa, xvi(4).

CONTENTS

VIII

INTRODUCTION

1. The increasing dependence of man on earth

Throughout human history man's dependence on the earth has
been growing through an ever stronger interlocking of
society with the earth as well as through the mere fact of
growth. An increasingly complex society, occupying most of
a planet with which it is so strongly connected, can expect
that its relationship to that planet will eventually become
one of its major concerns. Modern society has already reached
this point, as shown by the current difficulties and dislocations which have appeared in the areas of environment,
energy, and resources. For this reason the earth sciences
are beginning to assume an increasing importance, beyond
that which they already have. These sciences have at
present assumed considerable importance because of the role
they have played in the development and continuation of
modern society.

The fact, apparent at the present time, that the
earth is a major controlling factor in the development of
society is not new. Similar situations have arisen before
in which the earth has been a major influence in the

development of society as well as of science and the
attitudes of scientists. Seen in the light of history
the present-day manifestations of this long-term relationship of man to earth are not as inexplicable, frightening
and unmanageable as sometimes described. In broad outline,
the movement towards increasing dependence upon the earth
developed as follows.

The initial stages: man, a product and a user of the surface environment

Man, developing from his hominid predecessors during the
Pliocene epoch, (Table 1) was born as a part of the world
of mammals during the Cenozoic era (Table 1). At this time
mammals were the dominant form of animal life, having gone
through a rapid ascent throughout the Cenozoic era. The
latter, because of the profusion and vigour of mammalian life,
has been called "the age of mammals". The Pleistocene epoch
followed, bringing with it a period of profound environmental
change. This change, primarily in climate, produced major
dislocations in plant and animal communities and in the
conditions of human migration and settlement. Man began his
ascent as a creature of the earth at a juncture when a
great evolutionary thrust met with environmental challenge.
In response to the challenges of such surroundings, man
developed into a creature capable of great accomplishments.

T A B L E 1

Time Scale for the development of man

Cenozoic era	Quaternary period	Recent or Holocene epoch -12,000y	Neolithic age -8,000y
		Pleistocene epoch -2 my?	Paleolithic-Mesolithic age homo sapiens - 0.5 my?
	Tertiary period	Pliocene epoch -7 my?	development of man
		Miocene epoch -26 my	
		Oligocene epoch -37 my	primitive hominids
		Eocene epoch -53 my	
		Paleocene epoch -65 my	early primates spread of mammals

Mesozoic era

In this process he developed the ability to change the surface of the earth. By fire and axe he could clear the forests; by cropping he could affect the nutrient and hydrological cycles. These developments gave rise to the possibility of man-made exhaustion of the land, and of erosion and aridity. Often man in prehistoric times and after, by a predatory approach to his environment, used the land to the point of exhaustion and then moved on. On other occasions there were long periods in which careful husbandry led to a stable relationship with the earth. On still other occasions happy combinations of this sort ended and both land and cultures declined and died together. The two contradictory trends: the predatory approach and that of the husbandman continued side by side throughout history. In all these examples we see that agriculture and the surface environment were man's first links with the earth. Conditions changed as society developed and by late Neolithic times a new factor had entered into the relationship between man and earth. This factor was the discovery of metals and metallurgy, and it proved to be a major one.

Resources from within the earth

The discovery of metals and metallurgy represented a new bond between man and the earth. In prehistoric times

man learned to mine the ores from which metals could be
extracted. The utility, in industry and warfare, of
metals was such as to transform the relationships between
peoples, making possession of the sources of metals into an
advantage which was of prime importance in maintaining a
position of acceptable well-being. Metals, and later other
resources, became a major factor governing the fortunes of
peoples. Thus another bond between man and the earth, of importance comparable to that of environment and agriculture,
was forged. These factors were the principal ones governing
the relationship between the earth and human affairs. Destructive forces of nature: volcanism, earthquake, storm and
periods of deteriorating climate were also present and had
their effects. These forces were not, however, felt at the
time to the extent that they would be in later historical period.

The industrial age and closer ties with the earth

As industrial societies began to develop, the needs of transport
brought new forms of contact with the earth, first in the form
of canal and later railway and road construction and in heightened
interest in the oceans. At a later stage, air transport greatly
accelerated interest in the atmosphere, already generated in
connection with ocean transport and associated naval interests.
The development of communications, in the form of the telegraph
and later the telephone and radio, made the electrical phenomena

of the earth and its upper atmosphere a matter of importance.

Resources as well as transport and communications figured in the development of industrial societies. When the colonial empires of these societies were being built (and during the brief ascendancy of those empires) the world-wide survey and exploitation of resources heightened social dependence on knowledge of the earth and stimulated earth-directed science. Another aspect of these developments cannot be ignored. The aftermath of the negative social features of that period (imperialism, for which see Sec. 1.4) remains as a serious problem in present-day society.

As society developed beyond the mid-nineteenth century, finite limits to geographical expansion were confronted. At the same time society was becoming more complex. These circumstances led to a continuing growth of scientific interest in the earth. At the same time, these very circumstances also led to national and international crises. Such crises were habitually solved by the predatory approach, by war on neighbouring countries or by colonial expansion. When, in the late nineteenth century, all territories were essentially occupied by the industrialized world or its offshoots, the crises could not be relieved in the traditional ways. In the absence of powerful representatives of the opposite (husbandry) approach, major dislocations (national and imperial rivalry, culminating in World War I) resulted.

Vulnerability of modern society: total dependence on the earth

Crises of a similar origin have been asserting themselves periodically since the late nineteenth century. These dislocations, increasingly as time goes on, are related to environment and resources. It should not be forgotten that deficiencies in social structure have been aggravating factors in causing these difficulties to reach the proportions they have. In what follows it should be remembered that the solution of either type of problem (between earth and society, or within society) cannot be solved without the other. These are all part of a continuing series of misadjustments between society and the earth which has been developing for a long time. The present-day expressions of it can be understood and mastered only if seen as such. Panic at what are (falsely) viewed as uncontrollable difficulties can lead only to defeat. Equally difficult junctures in the development of society have been encountered before. At its best moments, science viewed such circumstances as a challenge. The depth and frequency of such crises are increasing with time, and science is now faced with some of the greatest challenges it has ever met. Let us hope that it meets them in a manner worthy of its greatest ages.

2. The earth and man: transformation or conservation

Man has always had to wrestle with nature and meet her
challenges. Whenever he has successfully done so,
whether through adapting to nature or changing his
environment, man has changed and advanced himself.
But the challenge today, as a result of the increasing
vulnerability of society and its dependence on the earth,
is extended in a new and radically different form, one
which has never been faced before (Sec. 1). A consequence
of this new situation is that the relationship between
man and nature has become a matter of widespread concern
in the present decade. Since such concerns have some
bearing on the earth sciences, it is important for us to
understand some of the forms which they have assumed.

Four main approaches can be distinguished. The
first advocates unbridled development of technological
society. This policy, which advocates development and
disregards any possible deleterious side-effects of
development which might result, seeks justification for
any such effects in the fact that development has occurred.
This approach is the present-day extension of the predatory
approach, already mentioned. It has produced acceptable
levels of development in the past, but its deleterious
side-effects now outweigh the advantages. This is so

because these side-effects affect the environment most of all, at a time when society has moved to a position in which its well being is strongly dependent on its relationship with the earth.

A second approach is one form of the classical socialist approach, in which again development is the central aim. However, in such a system, some of the worst side effects of development are reduced by planning and the recognition of many social and cultural needs. Such an approach has been of value in the early days of the Soviet Union and in present-day China, in leading to a much-needed acceleration in the rate of development. At the high point of development of these societies, "transformation of nature" was a phrase often heard. However, under present-day conditions, development alone cannot be the dominant requirement.

A third approach to man and the earth embodies the best of the humanistic tradition (including many elements of classical socialism), and some elements of human behaviour which go back to prehistoric times. This approach can be summarized by the title "The earth our garden" (CROWTHER, 1955). This is the modern-day extension of the husbandry approach, already mentioned, which has appeared as an alternative to the predatory approach all through human history. In this approach it is recognized

that man has, for example through agriculture, improved
the earth until its productivity has increased immeasurably.
This improvement has not been made at the expense of non-
developed nature. Agricultural man added to his fields
through constant attention and improvement. This development does not imply the destruction of nature: "Man is
a mammal and what happens to other species may well happen
to him. The most important of the many reasons for
preserving wildlife is that it may contain the key to
the survival of man himself...the teaching of philosophy
would be illuminated by an exploration of the truth that
nature and man are of one essence, and that the principles
of nature are enduring, and that to survive man must co-
operate with nature...(man)...must stop exploiting the
resources of the earth and begin to utilize them in a
rational, self-renewing system" (CROWTHER, 1948).

If these principles are followed, development of
the earth and society can continue far into the future.
Neglect of "the earth our garden", on the other hand,
means an eventual end to creative development of human
society.

The fourth approach, often heard today, calls
for the abandonment of any sort of development and for
"hands off" nature. This is an extreme approach which
comes from the anxieties of our times; for this reason

it will not survive the present era.

It is clear from the above review of ideas regarding society and the earth that there are a number of conflicting approaches to the conduct of affairs in this area. In spite of these differences, one idea is widely held: that if society is to progress in the future, a broad base of knowledge about the earth and its resources is required. Such a need is felt equally by advanced and by newly emergent nations. The human condition for a long time to come will be determined by how well this need is met. Not only has the increasing scale of man's activities on the earth brought us to this point, but also a critical turning point has been passed (Sec. 1.6) and now a scale of effort well beyond anything known before is necessary to give social development the momentum to carry on beyond this point.

3. History of science and its relevance to social development

The critical turning point now reached by society is one in which science is inextricably involved, both as a major factor which acted to shape society as it progressed towards that point and as a major factor required in finding a fruitful future course. As a consequence it may now be as important for scientists to take part in framing science policy as to produce science. This does not mean that a decrease in scientific effort should occur or that a moratorium on science should be declared, as has been suggested by some. It is in fact doubtful that the pace of science could be halted even if such advances were proven to be undesirable. This latter possibility is, however, unlikely and all signs point to the necessity of providing even more science (and what might be more important, statesmanship of science) to the major problems of society. Certainly earth sciences are an important part of the task of ensuring a fruitful future course for society because of the increasing dependence of society on the earth. Because of this position of earth science and of the place of history of science in the present-day context, the history of earth sciences has, beyond the intrinsic interest of the subject, an important place at the present time. It is in this context that the present book was written.

The growing understanding of the position of science in contemporary human affairs has caused the history of science, once considered by many scientists as a possibly interesting but peripheral activity, to be viewed in a new light. It is now becoming clear that a policy for science is a vital social concern. However, before science policy can be developed adequately, the factors which govern the development of science must be understood. In other words,

a "science of science" is required. The beginnings of
such a science can already be found in researches and
writings on the history of science. Those historians of
science who have related their studies to social evolution
have shown how science develops in conjunction with general
history, and that a study of the course of development of
a science is the key to its present characteristics and
potentiality for further development. For this reason, a
consideration of the history of his subject is rapidly becoming a necessary part of a scientist's education and
professional development. The present book is an attempt
to portray some branches of earth science in their relationship to history during decisive periods in the development
of the sciences of the earth.

Earth science and society

The earth sciences can play a special role in the solution
of some of the major social and economic problems. Ways
in which man can develop and maintain his society in
harmony with the physical environment and the physical
forces of nature (earthquakes or volcanic activity, for
example) depend on the earth sciences. Very many problems
of water, climate, soil and environmental control involve
earth science. Such problems are not purely technical
but are in part social, and therefore, cannot be solved

by science alone. On the other hand, they cannot be
solved by existing techniques without the application of
fundamental science: nothing less than the application
of science in a fundamental way is sufficient. Man can
learn to live with nature; also he can legitimately
aspire to improvements of nature such as the remaking
of the water regime and climates, or the controlling of
destructive elements of the environment by forecasting
and ultimately reducing their potency to harmless levels.
Thus we see that besides the history of earth sciences
themselves, their relationships to society and science
policy are important matters to consider.

4. The earth sciences

The term earth science itself is relatively new, reflecting
a stage in science which has been reached during the past
few decades. Sciences relating to the earth, which have
been in existence for some considerable time, have in
recent years become of special social interest and have
moved into the forefront of science. In conjunction with
these developments, a number of the disciplines engaged
in the study of the earth, along with some parts of space,
planetary, and solar science are beginning to move along
converging lines. It seems (Sec. 1.6) that an important

phase in the evolution of these sciences has been reached
and that a new, distinct science (distinct when viewed
against the background of the whole field of science),
broader than any of its component parts, is coming into
being.

The term <u>earth science</u> applies to a whole family
of sciences, a number of which are as follows: the
geological sciences; the geophysical sciences; the geochemical sciences; the oceanographical sciences; the
hydrological sciences; the atmospheric sciences, and the
geographical sciences.

It is usual (McGRAW-HILL, 1960) to group the first
three under a separate heading, the <u>geosciences</u>, to
signify that they are sciences of the solid earth. The
present book deals primarily with certain of the physical
geosciences, as is indicated by its title. There are, however,
many features in common among all the earth sciences.
Thus this more general term is used throughout the book
whenever it is felt that what is being said goes beyond
the particular group of sciences under study and applies
to earth science as a whole.

Scope of the book

Because a vast field in space and time as well as in
subject is offered by the earth sciences, a single study

such as the present book cannot hope to cover them all.
Limitations in subject were chosen as outlined in the
preceding section. As regards time, the book deals
mainly with the Scientific and Industrial Revolutions with
some reference to earlier and later periods. As regards
centres from which the developments came, those in Europe
are primarily treated because in the time period covered,
Europe was the main source of earth science. Some earth
science was initiated independently by former colonies
of European countries (for example the United States) or
by colonies with a strongly independent spirit (such as
Australia or Canada) during the period covered. However,
these developments did not come to the fore in earth science
until a later period, although there were a number of developments from these sources before that. It is hoped that
these developments can be covered in a sequel to the present
book.

5. Science policy and the earth sciences

Science policy

Interest in science policy has increased in recent years.
Much of this interest is as yet at a relatively uninformed
level, and fails to utilize what is already known about
science and society and their interaction. In spite of
the above-mentioned deficiencies, the interest is there,

and governments are turning more and more towards the
planning of science. Many of these planning decisions are
based on economics alone and fail to go beyond this limited
base. However, for an adequate understanding of the
development of science, a consideration of the wider
questions regarding science and society is required, and
is neglected in many of the current studies in science
and in much of the practice in science planning. It is
hoped that one contribution of the present book will be
in providing some understanding of these neglected factors
in at least some of the earth sciences.

The earth, science, and society

There has been relatively little understanding among
scientists, technologists, policy planners, teachers and
the public in general of how the earth sciences are
related to science as a whole and to man's relationship
with the earth.

Several developments in our century have made this
understanding very important. First of all, modern society
has developed in scale to the point where it must consciously
plan how best to occupy and develop the earth and its resources. In the past, similar problems occurred in
limited parts of the globe but there were always new areas
to move into (as the great migrations of the past: such as

those from Europe to the Americas, and from European Russia
to Siberia have shown). Under those conditions the problem
could be temporarily forgotten and left for future generations.
This can no longer be done, under the conditions of near-
full occupancy of the globe (Sec. 1.6). It is, therefore,
hoped that the book will be of value in pointing out the
significance of earth science to science and technology
in general, in light of these factors.

Chapter 1

THE DEVELOPMENT OF EARTH SCIENCE: BASIC PRINCIPLES AND PROCESSES

The earth is the ever-present background of all human activity. This background, acting through the natural environment, has played a part in all stages of the development of mankind. The development of human consciousness itself came in large part from the struggle of early man and his evolutionary forerunners with the natural environment. At this stage man was unaware of the fact that this environment, and therefore also man himself, is but a small part of a great system of nature extending far beyond the immediate surroundings. The awareness of a vast system came early in historical times, however, and is a part of the earliest writings of ancient civilizations. From this awareness came much of science, and the earth sciences are particularly closely connected with these ideas. Before reviewing the development of the earth sciences, let us examine science in general and the way in which science has developed throughout history.

1.1 The laws of growth of science

Science follows definite patterns of growth, and we
therefore may properly speak of laws of growth of science.
On the whole, growth is centred on a small number of
decisive periods and does not occur at a steady rate.
Growth is not, however, confined entirely to these brief
(although decisive) episodes of growth. The whole sequence
of growth forms a pattern which is of great significance in
the study of the evolution of science.

First of all, very few branches of science appear
without predecessors. There is usually a long period of
time in which the forerunners of a science develop slowly.
Then there is a rapid consolidation of the early forms of
the science and a sudden growth into its more developed
stages. This change is, in many cases, almost revolutionary.
We shall find that it very often accompanies social change
of some kind. In fact the history of science parallels
general history, which also develops with an uneven pattern
of growth.

There is no guarantee that once established, a
new science will continue to flourish. If it finds an
immediate application, perhaps to the technology or some
other vital concern of the age, it continues to grow,
while maintaining for some time the form it acquired

during the stage of rapid growth. If it does not find an
application it might stagnate or lie dormant for some time,
as, for example, the science of electricity once did.
Sciences may go through cycles of this type, reaching
progressively different stages, two or three times in
succession: that is, in escalation.

We can in fact map the growth of a science by a
sequence of growth curves. These growth curves are of a
type which develops in three stages. The curve rises slowly
at first, then swiftly through the peak period of development;
after this it may level off or even decline. When a second
cycle of development follows there may be a new rise beyond
this plateau, corresponding to a new development. If there
is a succession of cycles, we may find the resultant of
overlapping curves each representing a successive stage
of response of the science to historical development.

Logistic growth curves

It has been shown (PRICE, 1965) that a logistic
growth curve is a reasonable fit to some indices of the
growth of science. The logistic growth curve has been
shown to apply to a great many types of growth in
nature and in human society. DAVIS (1941) describes
the application of such a law of growth to biological
systems from very primitive ones to the most complex, as

well as to human population and to production data in studies of industry. The logistic curve grows from a slowly rising initial beginning, through a rapidly rising middle portion to a slowly rising final phase. This pattern of growth is a useful representation of the type of growth we have suggested for the earth sciences.

Characteristics of logistic growth curves

In our study we will utilize certain quantitative estimates of growth in an attempt to understand more fully the processes governing the development of earth sciences. These are such quantities as number of scientific papers, or a weighted index of the number of scientific discoveries, or indices of significant expeditions or of the other attempts to establish new knowledge of the earth. PRICE (1965) has demonstrated that simple indices of this type can tell a great deal about the development of science. He showed for the indices he used, that their cumulative total fits a logistic growth curve.

A logistic growth curve (y) rises from an initial phase of slow growth, through one of rapid rise centred on a point of maximum growth rate (called the critical point of the logistic) and finally moves into a phase of diminishing accumulation approaching an upper limit as shown in Figure 1.1. The slope, or derivative (y'

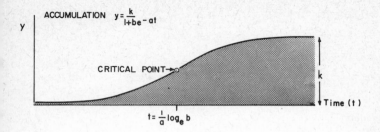

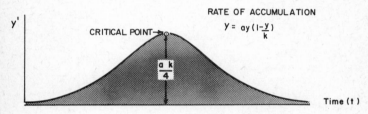

Fig. 1.1   Logistic growth curve and its rate of accumulation.

= dy/dt) of this curve is a symmetrical peaked curve with its maximum at the critical point of y. If y represents the cumulative total of a particular index, then the index itself is given approximately by dy/dt, and the peak of the index marks, very nearly, the critical point of the cumulative growth curve.

Examples of logistic growth curves in the earth sciences

We will see in subsequent chapters that a logistic curve
is a reasonable fit to indices of growth for a number of
disciplines in the earth sciences: for example, geomagnetism, mathematical theories of the figure of the earth,
and geographical discovery (Figures 3.1-3.4). Figure 3.1,
showing the growth of geographical discovery, affords a
particularly good example of growth curves. Geographical
discovery is in fact closely connected with the earth sciences
as we shall see in Chapter 2. Geomagnetism and geodesy first
developed in response to technical problems posed in the
extension of trade and navigation on a global scale, which
occurred in the fifteenth, sixteenth and seventeenth
centuries. Voyages and expeditions of geographical
discovery were at that time very much a part of the expansion in trade and navigation. Because of this
connection, an understanding of how the accumulation of
knowledge in geography took place is of help in understanding the development of earth science.

Figure 3.1 is, as will be explained in Chapter
3, an index of the extent to which the centres of geographical exploration were impelled to send out expeditions,
as it varied from the time of the first great European
explorations until the early twentieth century, when
virgin territory was virtually exhausted, and when

exploration began to merge with development. The Figure
can be analyzed into a sequence of logistic growth
curves in escalation, with the critical point of each
centred upon one of the principal time-divisions in the
history of science. Our analysis in later chapters will
show that relationships of this type enable us to understand the forces impelling the development of the earth
sciences.

1.2 A brief survey of the development of earth science up to the Scientific Revolution

Neolithic, ancient and medieval times

Man's relationship with the earth developed early, in
the form of mining and metallurgy. In Neolithic times,
extensive voyages and overland expeditions took place
(Figure 1.2), and there is evidence that prospecting
was one of the motivating factors (CLARK, 1969, p. 139).
The fact that prospecting could be successfully carried
out indicates an understanding of the geological environments of the ores being sought. There is, however, no
record as to whether or not this practical knowledge
led further to the understanding that a system of nature
lies beyond the relatively restricted world surrounding
the typically small settlement of the time. Such a
development may have taken place because in spite of

Fig. 1.2   Europe in the mid-second millennium B.C.
            from CLARK, J. G. D., 1969. World Pre-
            history: A New Outline (2nd Edition).
            Cambridge University Press, London, p.
            142.

their relative isolation in small communities, the people

of Neolithic times were far from being limited and

unimaginative creatures, but were highly productive
in technical developments and innovation.

At present, such ideas must remain speculative.
The first recorded accounts of man's awareness that the
immediate environment, and therefore that man also,
is but a part of a great system of nature extending
beyond the immediate surroundings came at a stage (in the
earliest civilizations) when astronomy had developed.
Astronomy developed in many of the early agricultural
civilizations as a means of determining time and seasons,
especially in those where timing in preparations for
planting and harvesting were critical. Astronomical
calculations are recorded as early as 2700 B.C.

Based on the resultant enlarged awareness of a
world beyond the immediate surroundings of the developing
civilizations, cosmologies soon developed. These new
ideas expanded in a variety of ways: they served as a
rationale for explaining and unifying astronomical
observations, and also as a projection of man's emerging
view of himself and the world. These elements varied
among the many scattered units of emerging civilizations;
as a consequence, many different cosmologies emerged. At
the same time intriguing connections are found among
cosmological ideas that emerged in widely separated places,
such as Babylon, Greece and China (NEEDHAM, 1959, v. 3, p. 212).

These cosmologies often appear fanciful to the modern observer who, unless he is a student of those societies, has not shared the view of the world giving rise to them.

For the earth sciences the important thing is that, regardless of the forms it assumed, the awareness of a far-reaching system of nature, by its very existence, had its effect and influenced the directions in which ideas were explored. One result was that geological theories were developed early in history. These theories show that the cosmologies of many of the ancient civilizations are based in part on an understanding of certain elements of geology and the geological history of the earth. The descriptions of these early theories given by CHARLES LYELL (1875, pp. 6-30) remain among the most perceptive in this regard. F. D. ADAMS (1938, pp. 8-50) details the many geological observations and inferences made in Greek and Roman classical antiquity. Many observations on mineralogy and on prospecting lore (including biogeochemical methods) are found in Chinese texts from the first century A.D. (NEEDHAM, 1959, pp. 591-680).

Another part of present-day earth science which flourished in classical times is geodesy. This branch of science developed in response to the need for maps in the far-flung ancient empires and their spheres of interest, as well as from problems which developed in astronomy.

Eratosthenes, who made the first recorded measurement of the earth's radius, was concerned with the problem of devising a projection to portray the known world of his (Alexandrian) times with reasonable accuracy on a plane map. He used this measurement as a part of the solution to this problem.

The earth as a sphere is an idea which arose early in many cultures. References to it are found in the fifth century B.C. in Greece, and in the first century A.D. in China. The radius of the sphere was estimated by Aristarchus of Samos in the third century B.C., and was measured with an accuracy comparable to that achieved in early modern times by Eratosthenes in the experiment described above, conducted about 200 B.C. Similar measurements were made in China in the eighth century A.D., following four centuries of preliminary work. These were repeated in the Islamic world in the ninth century, and again in Europe beginning in the 16th century.

Mining, one of the oldest industries, apparently contributed less to the broad vision of the earth (although the long prospecting expeditions already referred to may have contributed much more than we realize in this regard) than it did to metallurgy, and through the latter to chemistry. As regards the earth sciences, mining inspired the development of mineralogy. The forerunners of this branch of earth science date from classical times, as detailed by ADAMS (1938). Mineralogy was later in fact to develop early into a science with some characteristics of its modern form.

Another early development was that of the magnetic compass and its predecessors and the discovery of the directivity of the earth's magnetism. Even though references to the magnetic needle compass go back only to 1080 A.D. (in China), a lodestone spoon compass (apparently used on diviners' boards) was used in that country from certainly the first century A.D. and possibly from the second century B.C. (NEEDHAM, 1965a, v. 4, pt. 1, p. 229). The declination of the compass was discovered in China between the seventh and tenth centuries A.D., with recorded measurements dating from 720 A.D. The compass appeared in Europe about the twelfth century, with many features of it and its use suggesting its Chinese origin. The route of transmission is still unknown.

The Scientific Revolution

Economic and social developments in Europe turned these currents of earth science into new directions, which ultimately led to modern earth science. Technical, economic, and political developments in Europe culminated in the late fifteenth century in a great upsurge of industry and commerce, which transformed all phases of life including science. During this period, astronomical, navigational, mapping and geodetic techniques were extended across the globe in a predominantly maritime phase of expansion. One result of this expansion was that techniques (such as those for measurement of the earth's radius) which had existed for millennia, when extended through a range of latitudes, revealed new ideas about the earth. The discovery that the earth is ellipsoidal in shape rather than spherical led to a new era in geodesy which in turn inspired new ideas about the earth's inner constitution. This period marks the beginning of one of the geosciences (earth's shape and gravitation) in its modern form. At the time a spectacular upsurge of science as a whole occurred (the Scientific Revolution). In one direction of development, the great expansion of navigation and overseas trade created a demand for global maps, global astronomy, and navigational aids. In another direction, the growth of industry and commerce, with its demand for metals and other raw materials, stimulated lines of scientific work which were later to produce (as well as chemistry) the science of mineralogy. The principal ideas of Agricola, the "father of modern mineralogy" (ADAMS, 1938, p. 185) were presented in a book devoted

principally to mining, <u>De re metallica</u> (1556). Three
branches of earth science: mineralogy, geomagnetism,
and the earth's shape and gravity can be traced, in forms
recognizably close to their modern ones, back to the time
of the Scientific Revolution.

The fact that this period in history was one of
general change and development points to a close relationship between science and general affairs.

1.3 Scientific interest in the earth and its links with history

Nature of these links and an example of their occurrence

By and large, the developments in earth science, as in
all of science, have been a part of the broader movements
in general history. In later chapters we shall see
critical periods of development in which the earth sciences
acquired their principal characteristics, occurring
simultaneously with developments which are recognized
as being similarly critical in general history.
The links which must exist between the earth sciences and
general history, if these correspondences are more than
fortuitous, will be explored in some detail. These links
will be seen to be the primary influence on the development
of earth science. This does not mean that influences

outside of general affairs have not acted to form earth science (or any other science). Science has been formed by a whole complex of converging trends. The state of development of other sciences, the history and state of development of the science itself, the philosophical and general intellectual climate of the age, and the closeness or otherwise of science to technology are all factors determining the course of scientific development. All of these are intertwined, and the whole linked with general history. This link with general history, acting directly and indirectly, provides the broad framework of development, making possible a system of time-divisions of our subject matter. These divisions encompass a succession of scientific ideas, experiments, technical applications, and various kinds of fertilization from technology, all weaving a pattern of development of earth science, continuing from ancient times to the present.

As an example of links with general history we noted earlier that a measurement of the earth's radius was made in eighth century China. This was a major astronomical-geodetic undertaking involving measurements of sun-shadows and latitudes over a line 2500 km. long, carried out during the period 723-726 A.D. A detailed description of the project and its origins is given by NEEDHAM (1965a, v. 4, pt. 1, pp. 42-51). The work was undertaken to establish better the

relationship between distance and latitude, reflecting a
concern regarding the accurate determination of distance
and relative position within a large land mass. This problem had been recognized as a scientifically important
one for a long time in China, yet no such large-scale
measurement had been undertaken previously, in spite of the
fact that it had been technically possible for an equally
long time.

The general historical setting of the operation
is that of a period of re-consolidation of the large
land-based Chinese empire after a long period of frag-
mentation. Lines of communication were being re-established
over great distances. These circumstances, along with
the extensive interest in and contact with distant
countries, which was well-developed at the time, led
to a flourishing of cartography. Such a setting would
explain the upsurge of interest at this time in surveying
and the determination of distances between widely separated
points implicit in the fact that such a large geodetic
operation was undertaken. The general cultural climate
was favourable for intellectual pursuits since the empire
was enjoying a great cultural flowering (during the mid-
to-late T'ang dynasty, 618-907). Other developments with
potential application to surveying and mapping also occurred.
There was sufficient interest in the directive properties of

magnetized bodies to lead to the first recorded magnetic
declination, in 750 A.D. (NEEDHAM, 1965a, v. 4, pt. 1, p. 310
(Table 52) and SMITH AND NEEDHAM, 1967). The compass is an
instrument which is of great service in cartography. The
first measurement of declination was made by I-Hsing, "one
of the most outstanding mathematicians and astronomers of
his age" (NEEDHAM, 1965a, v. 4, pt. 1, p. 44). I-Hsing was
also the principal organizer of the geodetic traverse.
Here we see an example of a leading scientist turning to
the earth for fruitful problems, as had occurred elsewhere
before and was to occur afterwards. These studies of the
earth in T'ang times show a relationship between the society
and the science of that time, which is of the same sort that
we noted earlier as occurring in Europe during the Scientific
Revolution as well as continuing until the present day.
Thus this example shows that the relationship between science
and general history has remained very much the same over a
very long span of time.

The importance of the history of science

This particular connection shows, through the fact that a
similar pattern has been maintained for so long, that although the form and content of science changes rapidly and
the nature of society has been profoundly altered since those
times, there is an underlying link with general history

which as regards its mode of operation has changed very
little over the millennium which has elapsed since the time
of I-Hsing. If we believe that this link, which has already
survived profound changes, can reasonably be expected to
continue into the future then the study of such links is of
great importance. Thus the history of science is a necessary
part of understanding science and its role in the future of
our planet and its inhabitants.

1.4 Earth sciences in modern times

The basic driving force in the scientific study of the earth
and its links with society have changed very little over a
long time, and a systematic tracing of earth science as it
developed through the whole span of history is a very important task.

The present book, however, as an introductory
study, must remain confined to a more limited objective.
It was, therefore, decided that the most fruitful
starting point would be to examine some of the earth
sciences in their broad outlines and in their relationships one with the other, over the more limited span
of modern times. It was hoped that such a limitation
would allow a more concentrated view of the origins
of earth science as it is found today. It is important

to begin in this way because of the great need to
understand the place of earth science in our present-
day world and, more generally, to understand the
science of science necessary to understand and eventually
plan the future development of science and society.
We will then be concerned with growth stimulated by
the technological developments and problems that emanated
from Europe during the past four or five centuries.

Division into periods

The earth sciences as they are found today were shaped
in six decisive periods. The first was the Scientific
Revolution, already briefly described and to be treated
in more detail in Chapter 2. These six periods are
also important in general and economic history, and each
has left a mark on earth science as it exists today.
The fact that this mode of development occurred, in
successive bursts of growth, can lead us to an understanding of how these sciences can be integrated with
social development in the future.

The six periods (with dates rounded off to the
nearest decade unless some particular event begins or
terminates a period) are as follows:

(1) 1450-1700, the Scientific Revolution, corresponding
to the economic, political and cultural upsurge experienced

in Europe during this period;

(2)   1760-1830, the Industrial Revolution;

(3)   1830-1870, the mid-nineteenth century period of
European expansion;

(4)   1870-1918, the Age of Imperialism, when the easy
expansion of man's occupation of the globe, beginning
in prehistoric times, came to an end;

(5)   1918-1945, the first phase of the New Industrialism;

(6)   1945-the present, the second phase of the New
Industrialism.

Comparison with general histories of science,
such as that of BERNAL (1971), shows that these time-
divisions correspond to the principal periods of development of science as a whole.

Each one of these periods stimulated a distinctive
phase of growth in the earth sciences. We have mentioned
the Scientific Revolution and the three branches of earth
science that stemmed from it. Similarly, each subsequent
period had its characteristic effect on earth science. We
will begin with a brief description of the main episodes of
development.

The Industrial Revolution (1760-1830)

In this period, a radical change in the nature and scale of
industrial production occurred. Initially this great upsurge

of manufacturing created a demand for new technology related to energy, raw materials, and the construction of railways and canals. In this setting the geological sciences became established, essentially in modern form. The development of geology was the principal new direction given to the earth sciences by the Industrial Revolution. Of course the steady development of those branches of earth sciences which had been already established went on, and the forerunners of other branches continued or began to develop. Most important perhaps was the fact that the basic sciences of physics, chemistry and natural history went through significant changes in this period. These developments were to influence the earth sciences profoundly in later periods. At first, the developments of the Industrial Revolution were confined to Europe, particularly to Britain. The emphasis was on resources, manufacture, and consumption of the end products at home. Towards the end of the period, the emphasis shifted to the need for a world supply of materials and for a world market for the products. This new emphasis brought overseas transport and communications back as a primary concern, as they had been during the Scientific Revolution. As an example: it is interesting to note that in the scientific study of the sea, the period following 1775 was one of widening horizons, with a peak of activity reached in the interval 1815-1830 (DEACON, 1971, p. viii).

This renewed interest in the sea brought meteorology forward as an important study. Regular meteorological observations had been made in several countries since the late seventeenth century. The new interest in weather stimulated the development of the first weather map in 1826.

The mid-nineteenth century (1830-1870)

A new stage in economic life became apparent after 1830, especially in Britain, where industrialism had first been established. In this period, technical innovation was of lesser importance than it had been during the Industrial Revolution, and the expansion of existing types of industrial facility became the primary concern. The primary requirement of the growing industrial centres was for raw materials for their factories and markets for their products. In the technical sphere, interest shifted to transport and communications. This was the age of steamships, railways, and telegraphs, and was a period of strong belief in progress through the application of science. In this period, a new expansion of the branches of earth science already established was possible because the social stimuli for development of these sciences and because of the developments of science which had occurred during the Industrial Revolution. Methods of measuring intensity of magnetic field had been devised, and there was an upsurge of geomagnetic observations during

the period. Many of these developments accompanied the very
great activity in sea and land explorations, which reached
their peak in 1820-30. (Fig. 3.1). The concepts of evolution
and geological time were established as a branch of earth
science. International scientific organizations began to
come into being, such as the International Association of
Geodesy (1864). The name "geophysics" was coined in this
period, signifying the consolidation of those branches
which had already developed (previously referred to as
"terrestrial physics") into a distinctive field.

The Age of Imperialism (1870-1914)

The sharply increased international rivalry gave rise to
a new period in the nineteenth century: it is recognized by
historians as a distinct period with its own ideas,
economic background and style. A corresponding transition
can be noted in some branches of science, although there was
no break in continuity in other branches. A precise time
marking the beginning of this new period cannot be demarcated
for science as a whole. A new direction is more marked in
the sciences of the earth, as might be expected, because the
earth is closely connected with the cause of the historical
change. A good example is afforded by the growth of geomagnetic
observatories (Figure 1.3). The period after 1870 was one of
international co-operation in obtaining scientific data of a

Total number of geomagnetic observatories in operation

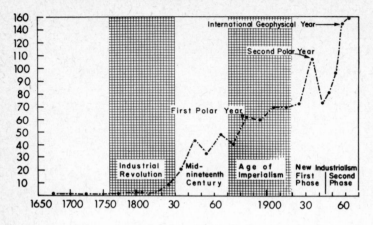

Fig. 1.3   Accumulation of number of geomagnetic
            observatories

type which could not be obtained by any one nation. The International Meteorological Organization was founded (1878) in
this period, and international projects of a new type, beginning with the First Polar year were organized and carried
out. In this period there began a series of great oceanographic
voyages such as those of the Challenger (1873-6), Gazelle
(1874-6), Fram (1893-6) and Valdavia (1898-9). The end of this
period marked the end of a period in science in general.
BERNAL (1965, p. 518) characterizes the end of this period as
the end of a "heroic" or "amateur" stage in modern physics,

where individual achievement following a revolution in physics
(which occurred in 1895) led to a whole new world of science.
This new world included the quantum theory, the structure of
crystals, the Rutherford-Bohr atom, and the theory of
relativity. As regards sciences of the earth, the theory of
Continental Drift was first proposed in this period by Alfred
Wegener. It is significant that he worked primarily as a
scientist in Arctic exploration. His work was characteristic
of the Polar Year enterprises.

The New Industrialism, first phase (1918-45)

Science entered a new phase following World War I. In this
period there occurred the "first large scale entry of industrial techniques and organization into physical science"
(BERNAL 1965, p. 518). This development transformed virtually
all sciences, and the transformation was motivated by the
same developments which set a new era for the earth sciences.
The new world of physics included radio, television and control mechanisms. Scientific discoveries included neutrons,
positrons, mesons, cosmic rays, the ionosphere, and signals
from radio stars. The electron microscope was an important
invention in this period, allowing new understanding of the
structure of matter.

This was also a new period in earth sciences because
following the Industrial Revolution industry gradually changed

its character from the traditional reliance on coal and steam
to an increasing application of new energy sources and new
materials. These sources and materials became much sought after
in the face of the growing demands of industry in the present
century. As a result, geological, geochemical, and geophysical
prospecting grew rapidly and new methods of mapping and exploration began to develop. In the course of this development,
geology and geophysics grew and changed radically.

International co-operation continued, in the Second
Polar year, because of the awareness of scientists that the
further understanding of geophysical phenomena in geomagnetism,
meteorology, aurorae and other branches of geoelectricity would
advance science as well as practical applications such as to
air and sea navigation, weather forecasting, and radio communication. Because of the worldwide economic depression
following 1929, there were members of the scientific community
who advocated indefinite postponement of the project.
Fortunately, enough supporters for continuation were found;
the project went ahead as planned, in 1932-33. It is interesting
to note that the general difficulties of the times prevailed
as regards the number of continuing observatories. The upsurge
in numbers of observatories generated by the project died
off immediately after. The scientific results were of
lasting value, however, but the lack of lasting growth in
observatory facilities reflects the general social background,

dominated by economic crisis and then war (Figure 1.3).

The advent of aviation had created a new incentive for understanding the atmosphere. The development of radio heightened interest in ionospheric science and, consequently, in related sciences: geomagnetism, geoelectricity, and aeronomy. In this period, science and technology were far in advance of that available in earlier periods, although not on the qualitatively new level that was to be available in the following period. For example: radio was available, but still not reliable enough to relay masses of data from remote locations as was to be possible in telemetry of data in the later period. At this stage radio provided problems which set the stage for research, but it did not yet provide a revolutionizing means of research, as it was to do later.

This period was, because of the new science available, one of rapid accumulation of new ideas in earth science. However, a revolution in earth science comparable to that in physics in the preceding century could not occur at that time because the basic data were not yet available. They did not appear until a later stage of science involving efforts on a national scale came into being. There were by this time many nationally sponsored efforts, and internationally organized projects, but not on the scale that was to occur in the following

period. After 1945, a new phase began in general
science, characterized by BERNAL (1965, p. 518) as one
of "government science". In this period De Solla
Price's transition to "big science" (PRICE, 1965) occurred.
Both the incentive for growth and the means of achieving
it came into being for earth science.

The New Industrialism, second phase (since 1945)

The development of atomic power, rockets, electronics
and a new, powerful basic science meant a new scale of
developments in earth science. The International
Geophysical Year catalyzed space exploration, and because
of the new level of basic science studies of the sea
(elevated to a position of great interest and importance
by largely social concerns) have led to an understanding
of the basic forces controlling the evolution of the
earth. The general scientific background has provided
radar, atomic science, radio astronomy, electronic
computers, sophisticated control systems, advanced
telemetry, transistors, and lasers.

The oceans were at last in this period thoroughly
explored and represent the last earthbound frontier.
The new methods of an advanced science, because of their
availability, made it possible to carry out what nations
would have liked to do in earlier periods such as the

Age of Imperialism. But the means were lacking (and perhaps the vision of what could be done was impossible at that time) in these earlier periods. One feature that is retained is that of international scientific co-operation, with fierce nationalism hidden underneath.

Thus the period involves the old rivalries, coupled with a scientific capability and a scientific vision many times more powerful than in previous periods. Scientific scrutiny thus had finally spread effectively over the whole face of the globe. The main clues in the discovery of the motive power and mode of evolution of the outer parts of the earth lay in the oceans, the geoscience aspects of which had been largely unknown to science previously. Therefore, in this most recent period a new phase of earth science, representing a major advance in scientific knowledge, has been established. The fact that this development did not occur until this period shows that the influences other than general history previously discussed as determinants of the history of science (Sec. 1.3) are necessary to explain the development of earth science. Some technical developments that made the difference were: continuously recording sensors for a wide range of physical effects (magnetism and seismic waves to mention only two), with telemetry if necessary; and the capability for sophisticated analog and digital processing of signals.

The latter has made large-scale data acquisition possible, along with a significant advance of the geosciences on the scale of sophistication (Section 1.5). Thus the second phase of the New Industrialism involves not only new energy but new scientific techniques which are high on the scale of sophistication (Section 1.5). Also techniques taken from the mineral exploration industry have been important, such as deep sea drilling, deep sea echo sounding and seismic display techniques which allow the portrayal of sea-floor sediment and other layers beneath the sea bottom, as well as remote-control electronic packages (making, for example, deep sea mass-produced heat flow measurements possible). The steps into space brought phenomena in the upper atmosphere and in space into man's environment; they became, therefore, something more than remotely located effects as they had been previously. Lunar geology and geophysics on the the moon are an important new feature of this period.

1.5 Characteristics of the earth sciences

The scientific study of the earth possesses special characteristics because of the place of the earth in nature. This special character of earth science has determined the way in which it developed and its place

and importance in science. These characteristics are
sufficiently unique to require a special philosophy
of the earth sciences. There are reasons to believe
that this philosophy will have great significance in
future developments of science. This special nature
of earth science arises from the way the earth fits
into the scale of natural systems.

The scale of material systems in nature

The earth occupies an intermediate position in size,
when we consider the scale on which natural systems occur.
When scale is the factor being considered, the degree
of aggregation of matter in any natural system is the
prime factor in determining its relationship to the rest
of nature.

In a survey of nature, when we consider the
systems into which matter is organized, we find a sequence
of size that runs from subatomic particles through to
cosmic bodies. We can, in this sequence, discern a
succession of levels in scale. We shall find that the
earth occupies an intermediate position in this system
of levels, and that this position not only leads to a
complex relationship between the earth and science, but
also elevates the sciences of the earth to a position
of great importance in science.

Division into three levels

We may consider the following sequence, which divides
aggregations of matter and their associated natural
phenomena into three levels: microscopic, macroscopic
and cosmic. The microscopic-macroscopic division is
recognized as an important one in the philosophy of
natural sciences, and has been much discussed in the
philosophy of physics. A three-fold division similar
to that outlined above has been proposed on various
occasions in discussions of the earth sciences, for example
by VAN BEMMELEN (1967) and J. TUZO WILSON (1966). The
consequences of this fact of nature, that natural systems
fall into such a classification, are many. First, let us
establish the basis of the classification. A classification
of this kind follows naturally when matter and the scale
on which it is organized are taken as the bases of classification.

The microscopic level is concerned with the
individual behaviour and properties of the elementary
particles which constitute all materials.

The macroscopic level is that of observations on
a grosser scale where averaged effects of the elementary
particles are observed, such as might be the case for
experiments in an ordinary terrestrial laboratory.
We are commonly concerned in this case with matter in

a more highly organized form than is found at the microscopic level, and therefore with large numbers of elementary
particles arranged in some system, such as a crystalline
lattice. In the example of a lattice, interactions can
occur between particles, because of their ordering in
proximity to one another, giving rise to new properties
of matter, hence to new phenomena which do not occur
on the microscopic scale. These properties and phenomena
are, therefore, characteristic of the macroscopic level.
More complex systems occur and include living things
and the communities they form on the earth's surface,
giving rise to additional phenomena characteristic of
the macroscopic scale. We may perhaps even include
some systems arising within human society, such as those
engineering systems (dams and irrigation, for example)
which are interlocked with the natural environment to
such a degree that they may be treated for some purposes
as part of nature, and therefore as subject to our
classification. It should be understood that in our
usage the term macroscopic scale is not intended to
extend in scope beyond that of ordinary laboratory
experiments or of technology and industry.

On the cosmic scale, among the systems we may
note first are those that would be studied when a body
such as the earth or another of the planets is considered

as a whole, or in substantial part. The scale for planets
as a whole is in contrast to more limited portions of
these bodies such as individual physiographic features
or a biological community on Earth. The planets
thus treated find a place among the grander systems of
astronomy such as the sun, stars, galaxies, and bodies
of intergalactic, interstellar, and interplanetary
matter.

Natural laws

That the system of levels just described represent a
scientifically valid and objective classification is
shown by the fact that systems on each level exhibit
characteristic phenomena, which can be described by a
corresponding set of natural laws, and by the fact that
these phenomena and laws are in the main unique to the
level in which they occur.

As an example, taking a case on the cosmic scale,
the distribution of the chemical elements throughout
the earth is governed by a particular set of laws
peculiar to the chemical nature of the earth, those
of geochemistry. These laws were originally formulated
in the course of scientific studies of the earth.
However, these geochemical laws exhibit another feature:
they can also be deduced from the laws of chemistry as

formulated and applied on the macroscopic scale, in ordinary laboratory chemistry. The basic laws of chemistry had been established before the advent of geochemistry. Once the latter had been established, it was not long before the connection between the two was recognized. It is unlikely, considering analogous cases, that this connection would have been recognized even though the scientific basis had been long in existence, until chemical phenomena on the earth-scale had been recognized. The chain of development of ideas led even further, eventually to cosmochemistry, a science which has made fundamental contributions to the understanding of the universe.

In comparing these two levels of natural laws we note another feature of them. These laws on the higher level can be deduced from those of the lower (the macroscopic) level, but not vice versa. The phenomena of the higher level represent a new complexity of matter and cannot occur on the lower level. There is a one-way dependence which cannot be reversed. The natural systems to which the geochemical laws apply cannot be found on the laboratory or industrial scale because the latter are simply too small in extent and they support processes with lifetimes that are too short to allow geochemical phenomena to come into play. These relationships are,

in this example, the result of the evolution of the earth.
Materials to which the macroscopic laws apply when they
are confined to small samples have been organized into a
system on the cosmic scale, giving rise to phenomena which are
impossible on the lower level and unique to the higher
one.

A survey in similar fashion of a great variety
of systems, along with their associated phenomena and
laws establishes an ascending order of scale (microscopic, macroscopic, and cosmic) as a valid classification
in nature.

Each level has its own systems which in turn have
their corresponding sets of natural laws, all characteristic
of that level. The process of evolution of the earth
has in general taken systems on the microscopic and
macroscopic levels, and formed them into systems on a
higher level of scale. An important feature of the
earth and planets is that they lie near the border
between macroscopic and cosmic in scale. This fact
has a bearing on the nature of earth science, which will
be noted later. Levels of scale form a particularly
valuable concept, which is of service in tracing man's
expanding vision of nature, as it unfolded through
history. It affords as well a means of understanding
the relationship between terrestrial and cosmic sciences.

An expanding sphere of action of mankind moving into successively higher levels of scale, carrying man's understanding of nature from the narrow, local environment to the terrestrial and finally the cosmic scale, is a basic feature of the course of development of scientific studies of the earth.

The bearing of levels of scale on scientific theory

Because of the scale of phenomena studied by them, the earth and astronomical sciences have been of fundamental importance in science. This importance may be seen by examining the nature of scientific theory, bearing in mind the fact that scientific theory is ultimately derived from experiment. An important feature of natural law is the fact that it is approximate. The approximate nature of natural law comes about because man must explore nature piece by piece, gradually extending his knowledge. The precision of this knowledge is limited by the accuracy and resolving power with which he is able to view nature at a particular stage in science. At an early stage, the resolving power of scientific observations is not great enough to distinguish among a number of possible explanations of the observed phenomena; thus initially, several theories may serve equally well. It is usually assumed that to advance beyond this stage science must await increasing precision of observations, so that it can decide among the various theories which were, in the earlier stage, equally probably (DUHEM, 1954, pp. 171-172).

However, historically, it has been found that another means for advance sometimes presents itself, often leading to far-reaching and profound theory. By this means observation is extended on a broader scale in time and space, making scientific advance possible without awaiting greater precision of laboratory experiments to improve on natural laws as known at a particular time. It may become possible, by means of this wider perspective, to separate the various possibilities suggested by the more restricted terrestrial experiments, thereby finding a more widely applicable theory.

In this view, scientific observation and experiment on the cosmic scale (with the latter including the study of the earth) has had an important place, along with terrestrial observation and experiment, in the development of science.

The Copernican Principle and the link between the earth and science

The earth has been, time and time again, the testing

ground of scientific theory. Often this has been so because the earth provides features for study that are on the cosmic scale. The relationship between these and laboratory science involves the same relationship between laboratory and cosmological problems. Thus this relationship is important to discuss here. This interesting meeting point of earth sciences, basic science, and cosmology deserves discussion. These connections are of utmost importance for science in the future, and we begin by examining cosmology and its bearing on scientific knowledge. The alternation that has taken place between earth-bound experiments and the cosmic realm in the development of physical theory is based on the assumption that a determinible relationship exists between physical law on the earth's surface and that in the cosmos. When the full sweep of time and space in the universe is considered, some schools of cosmology question this assumption, but for a more limited region and span of time, virtually all are agreed on what has been termed the Copernican principle, which states that within the compass of the solar system and during the time involved in the evolution of that system, the same physical laws hold everywhere as for a terrestrial experiment. The principle is widely held as being true, even if only because the observations of planetary and solar astronomy are all satisfactorily explained by principles

derived from terrestrial science. In addition, the fact
that terrestrial experiments can be repeated even though
the earth is moving and is thereby changing its position
between experiments is further evidence of the Copernican
principle (BONDI, 1961, p. 11).

There are some cosmologists who hold that physical
laws vary with position in the universe and with the age of
the universe. As expressed by DE SITTER (1932, p. 307)
"from the physical point of view everything that is outside
our neighbourhood is pure extrapolation, and we are entirely
free to make this extrapolation as we please, to suit our
philosophical or aesthetical predilections - or prejudices".
An age parameter as one of the determining quantities of the
constant of gravitation in the cosmological system of A. E.
Milne is another example of this sort of view.

Even these critics agree, however, that the Copernican
principle is at least approximately true, especially when
applied to "our neighbourhood". DE SITTER (1932, p. 307)
states, regarding the danger of large extrapolations from
terrestrial experience, that "it is only within the solar
system that our empirical knowledge of the quantities determining the state of the universe...extends to the second
order of smallness".

It is of course also important to remember that, as
far as a direct experience is concerned, it is also an extra-

polation to say that this uniformity of physical law holds
on a time scale longer than that of human history, even within
the solar system. However, other cosmologists, such as H. BONDI
and T. GOLD (1948) believe that in principle terrestrial
experience can be validly extrapolated. As stated by Bondi:
"apart from local irregularities, the universe presents the
same aspect from any place and any time" (BONDI, 1961, p. 12).
In this view, it is scientifically valid to "attempt to apply
our scientific knowledge to distant astronomical objects and
to the ancient geological history of the earth and the solar
system" (BONDI, 1961, p. 12). These arguments were put forward in support of what its authors refer to as the perfect
cosmological principle, formulated by BONDI and GOLD (1948).
This principle is an extension of the Copernican principle,
which in turn was a return to cosmological ideas held in
Greece in pre-Socratic times.

That the properties of atoms explain the earth, sun,
moon, and stars equally well was central in the cosmology of
Democritus (FARRINGTON, 1936, p. 85). This view was later
eclipsed, and in Aristotle's plan of the universe, in which
the celestial spheres were regarded as being fundamentally
different and isolated from the terrestrial spheres (FARRING-
TON, 1936, p. 148). Subsequently, belief in the "Aristotelian
distinction between celestial and terrestrial matter" remained
the dominant philosophy until some forty years after the

publication of Copernicus' _De Revolutionibus orbium caelestium_ (1543) (BLAKE, DUCASSE and MADDEN, 1960, p. 25). This period marked the beginning of a return to the pre-Aristotelian view. Notable in this movement were Copernicus, Kepler, and Galileo (BLAKE, DUCASSE and MADDEN, 1960, pp. 41-49). In his _Epitome astronomiae Copernicae_ (1618), "Kepler demanded that the same laws that apply to terrestrial physics must be applicable also to celestial physics -- but the latter must not be developed from independent hypotheses" (quoted in BLAKE, DUCASSE and MADDEN, 1960, p. 39). Further developments of these emerging ideas were to have great significance for the study of the earth two centuries later. Astronomy was, by virtue of these ideas, stripped of its licence to construct hypotheses which fitted astronomical observation but which did not satisfy further requirements, such as the laws of physics. In this new setting, many speculative systems were eliminated, reducing different possible constructions on astronomical observation but which did not satisfy further requirements, such as the laws of physics. In this new setting, many speculative systems were eliminated, reducing different possible constructions on astronomical grounds alone often to one. In his explanation of why the Ptolmaic hypothesis did not have the same consequences as the Copernican, Kepler remarked, "different hypotheses have not the same consequences as often the less skilled suppose" (quoted in BLAKE, DUCASSE and MADDEN, 1960, p. 41).

Scientific observation and experiment, their relative degree of sophistication

In studies of the earth, we find a whole range of sophistication in the type of data produced. The differences in effectiveness among these various types of study are of importance in determining whether or not a really productive level of science has been reached.

We may think of a scale, running from simple observation to full scale controlled experiment. The degree of sophistication of a science may be judged by its position on this scale. As a science develops and advances, it moves upward on the scale.

By observation we mean, in its simplest form, the mere description of, or the measurement of parameters related to, phenomena of (in Francis Bacon's words) "uncontrolled nature". We can imagine a whole sequence of types of observation grading from the simplest through more sophisticated designed observations to true experiments, such as the rigidly controlled experiments of modern physics or chemistry. The term experiment may be applied also to the more sophisticated types of designed observation, lying on the upper parts of our scale. As an example, of sophisticated designed observation we may note the arrays of seismological stations that have, in recent years, been emplaced in a manner so as to delineate selected zones of the earth and selected portions of its interior with maximum effectiveness. These have contributed much to the recent surge of understanding of the earth and the processes driving its development, mentioned in an earlier section (1.4).

In general, advance in science has consisted of devising observations and experiments successively higher on the scale of sophistication. This development was foreseen by Francis Bacon, who held true experiment in high regard. He expressed the view that real advance in science requires the "facts of deliberate experiment, rather than of the observation of uncontrolled nature", and that "when by art and the hand of man she (nature) is forced out of her natural state, and squeezed and moulded, and strained and vexed, she betrays her secrets more readily than in her natural freedom" (from The Great Instauration, quoted by BLAKE, DUCASSE and MADDEN, 1960, pp. 51 and 54).

Because of the nature of the problem, sciences of the earth were obliged in their early stages to be content with mere observations of the constitution and behaviour of the earth. The development of certain branches of geophysics, outlined in Section 1.3, was the beginning, in the earth sciences, of the movement upwards on the scale of sophistication. The first

rapid elevation on the scale occurred when geophysics
began to develop. At this point, designed collaboration
and experiment and sophisticated data reduction techniques began to elevate its observations significantly
on the scale of sophistication. The synoptic world-
wide collaboration of the Göttigen Magnetic Union or
of the First and Second Polar Years and, later, the
application of advanced data processing techniques
capable of separating the effects of processes with
different time-scales led to a significant upgrading
beyond simple observation. At this stage we are seeing
(as for example also occurred in astronomy in the
development of astrophysics) an application of the
techniques of measurement developed in physics applied
to the observation of nature on the cosmic scale. It
should be noted that even though geophysics and astrophysics have one feature in common (application of
physics on the cosmic scale), there is one important
difference. Geophysics is higher on the scale of
sophistication because the degree of control of observation is greater. This fact has considerable
significance in present-day science because geophysics,
with its possibility of design and control of observation,
is being extended by means of space probes to the study of the
planets. This represents a new departure for science, the

possibility of true experiment in astronomy.

A further development came with the International Geophysical Year, when the first long-term programmes of research from space vehicles were begun. It then became possible to design trajectories of these vehicles so as to obtain systematic coverage of cosmic bodies beyond the earth by means of remote sensors and, in some cases, by direct sampling. It became possible for the first time to carry the methods of true experiment beyond the earth.

In the space age which thus began, the methods of sophisticated observation which originally arose in geophysics are being extended to planetary and astronomical science. These methods applied in these new realms may lead to a new, higher stage in scientific technique.

Sciences of the earth, planets and space
- an approaching synthesis?

Scientific measurements by space probes in the solar system are bringing the geophysical type of experimentation directly to cosmic regions. In technique and scientific content, these studies are an extension of earth science. They have, beginning with the International Geophysical Year and the endeavours from which it originated, evolved

from earth science through the application of techniques and theories developed there. In addition,
all of the bodies in the solar system have a common
origin and therefore are all amenable to similar methods
of scientific study. Thus planetary astronomy, earth
science, and the new planetary science now being developed
in the course of experiments in space and on objects
in the solar system, are drawing together into what
appears to be a new more widely embracing yet distinct
area of science. This new combination almost certainly
will supersede all previous methods of studying the
solar system; consequently, under such a strong stimulus
for development we may see a synthesis of most or all
of the space, planetary and earth sciences in the
future. The achievements of geophysics are among the
first examples of the power of refined experimental
methods when applied to a cosmic body. In the new
setting of the planets and space this approach will
undoubtedly mean the revelation of a new dimension of
knowledge and understanding.

1.6 The earth and science

We have examined some of the basic characteristics of
earth science. We will see in later chapters how these

characteristics have acted to determine the way in
which these sciences have developed. In addition, these
characteristics have led to an important feature of the
scientific study of the earth, to the fact that the earth
has repreatedly figured in the most fundamental of
scientific quests, that of establishing the basic laws
of matter and energy. In magnetism the earth figured
in the work of William Gilbert and Carl Friedrich Gauss;
in gravitation the earth figured in the work of Isaac
Newton, Pierre Simon Laplace, and Albert Einstein. The
reason that the earth appeared at these critical points
in scientific development follows from the fact that
natural laws are based ultimately on the results of
experiment. By far the greatest amount of scientific
experiment has, throughout the history of science, been
done in laboratories and within at most the life-span of
individual observers. But natural laws ultimately describe
processes operating on a scale in time and space appropriate
to the universe as a whole. Thus small-scale experiments
cannot in general lead to more than an approximate
understanding of fundamental principles; for the
more fundamental advances, a broader scale is required.

When we go beyond terrestrial experience and
consider bodies such as the earth or the planets, or
the larger cosmic bodies, we meet new phenomena, charac-

teristic of aggregations of matter on a vast scale, showing more clearly the fundamental natural laws. Perhaps the most striking example is the discovery of the theory of relativity, which owes its existence in part to the Michelson-Morley experiment using the earth as an observational platform.

The earth is related to fundamental science in a complex way. An understanding of this relationship leads us to an understanding of many features of earth science; perhaps more important is the fact that earth science is merging with other sciences into a new science, planetary space science. Thus, through earth science this new science has a strong connection as a generator of fundamental science. Before we discuss this development, let us examine earlier examples of the earth in the development of science.

Newton's law of gravitation and the earth

Isaac Newton deduced his law of gravitation originally from planetary motion. The problems which led scientists of the period to search for such a law were not only in the field of astronomy, but were also to a large extent bound up with interest in the shape of the earth, stimulated by the needs of navigation. Newton's proof of his law of gravitation was not confined to astronomy,

but to a considerable extent involved application to the
earth. This was in the Copernican spirit, applying a
theory originally derived in astronomy to the earth and
its parts. He dealt with problems such as the earth's
shape, the mutual gravitation of its parts, and the
effect of external bodies on it. Some headings in
Book III of the Principia are: "Problem...to find and
compare together the weights of bodies in the different
regions of the earth"; "proposition...That the flux
and reflux of the sea arise from the actions of the sun
and the moon"; "problem...to find the force of the sun
to move the sea" (NEWTON, 1687). These applications
occupied an important place in establishing the theory
of gravitation for the following reasons. The earth
offered the possibility of observation at close hand;
and on the earth, gravitational effects occurred
sufficiently strongly to allow a clear view of such
phenomena. For these reasons, applications to the earth
had a decisive value because they brought to light
phenomena which were somewhat different from those of
planetary motion from which the theory was originally
derived, while presenting clear and measurable data for
analysis. It is significant that Newton himself placed
very great importance on the earth as a proving-ground
and as a place of application for his newly developed
theory.

The theory of relativity

The earth had an important influence on this particular development of science. One of the basic features of scientific theory and its development (levels of scale) is involved particularly strongly here. A striking example of the discovery of new, basic principles (and ones which were previously unsuspected) when experiment moves upwards in level of scale is the discovery by Albert Einstein of relativistic dynamics. "The starting point of Einstein's thought was the difficulties inherent in a branch of nineteenth-century physics; the attempt to generalize the electromagnetic theory of light by showing that the apparent velocity of light was dependent on the rate at which the observer travelled through the supposedly fixed ether. This was the celebrated Michelson-Morley experiment, the greatest negative experiment in the history of science" (BERNAL, 1965, p. 530). In this experiment the earth acted as platform, the only one then available, carrying the apparatus for a sophisticated laboratory experiment at a sufficiently large velocity to allow a great enough range of velocities and the corresponding effects on the travel-times of light to be measured. These results pointed to a whole new view of dynamics which became clear once motion was viewed on this larger scale. "No difference

whatever was found in the velocity of light at whatever
speed or in whatever direction the observer was moving"
(BERNAL, 1965, p.530). This effect was explained by
Einstein's special theory of relativity. The subsequent
history of the theory of relativity is equally striking.
Einstein's general theory of relativity predicted
certain effects of gravitational fields on the trajectories
of light and material particles. These effects are
small and most easily observed by their cumulative
effects on the huge trajectories of cosmic bodies. The
effects of the mass of the sun on the trajectories of
light beams from the stars and details of the orbit of
the planet Mercury were predicted by the theory to be
observable by astronomical means. These observations
were made soon after their possibility was predicted,
giving experimental support for the theory.

It is of interest to note that in later years
there have been some attempts to deny a connection between
the Michelson-Morley experiment and the origin of the
theory of relativity (POLANYI, 1958, p. 10). It would
appear, however, from Einstein's earlier writings that
the experiment did play a role in his development of the
theory, and perhaps most important of all, in the way
he chose to present the idea to the scientific world
(GRÜNBAUM, 1961, p. 50; EINSTEIN, 1952, p. 37).

Space and planetary sciences

We noted in an earlier section that earth and space sciences are at the present time merging into what may be a new, broader science. The reason for this development may be seen by examining the way in which the parent streams of the new science have developed.

The scientific study of the earth has by and large served scientific or social needs, some of them ranking among the foremost problems of the day. The development of reliable methods of navigation on long voyages at sea was a foremost technical problem in the two centuries following 1500. The earth's shape and gravity, as well as geomagnetism, figured largely in the solution of these problems. The systematic search for raw materials for manufacture, and the need for still greater improvement in sea transport and communications involved in the sciences of 'the earth in yet another way in the century following the Industrial Revolution. These were among the developments which stimulated the formation of the earth sciences as such, and gave impetus and direction to their subsequent growth.

The separate sciences of the earth began merging into earth science as the result of a fundamental change in man's relationship to the earth, which took place the the late nineteenth century.

Two features of the new relationship were responsible for the onset of the new development within science as applied to the earth. One feature was the fact that man's expansion over the face of the globe, at least in the ways he had known since his earliest development, had essentially come to an end. He could no longer simply move on or expand his territory whenever the supply of land and resources needed by him diminished in relation to the demands placed on his society by an increasing sophistication and a growing population.

The second feature was the fact that the dominant societies had become highly sophisticated and showed signs of becoming even more so, making them highly dependent on their relationships with the earth and with each other. In fact, dependence has turned into vulnerability a century later. Other factors were linked with the new sophistication and interdependence of the industrial centres of the late nineteenth century. Rapid overseas journeys and rapid communications were of utmost importance. These demands on technology led to a renewed interest, on a scale hardly matched since the Scientific Revolution, in geomagnetism. Also (because of the interference from natural electric phenomena which was encountered in telegraphs and submarine telegraphy) an interest in geoelectricity was stimulated. Widespread control of mineral and other natural resources became necessary for the developed societies if they were to maintain their growing industries. In this setting, calling for international interest in resources and in the solution of global geophysical problems, international co-ordination of effort came to be seen as a necessity. Such a need, for planning and organization, even if in some cases only dimly perceived or not perceived at all, was widely felt. It was felt widely and deeply enough that a dozen or more nations provided support for co-ordinated scientific expeditions to the Arctic during the First Polar Year in 1882. This was a pioneering (although not the first) effort in international scientific collaboration in the study of the earth.

It has been pointed out in preceding paragraphs that the desire for geographical expansion played a key role in determining the nature of major geophysical efforts such as the First Polar Year. The importance of this factor is shown by the fact that an explorer, Karl Weyprecht, whose attempts to operate as such were forced into new channels by the new situation, was the originator of the First Polar Year (HEATHCOTE and ARMITAGE, 1959). These directions in earth science gave rise to the consolidation of what we may call the international geophysical tradition. This feature of geophysics has continued to the present day, and its effect on earth science has intensified.

The first entry into space was made by a military rocket, the V2, in 1942. Military experiments with rockets continued through the 1940's and 1950's. These experiments contributed to rocket technology which soon was to be diverted into scientific channels. Rockets were also used in this period to carry sensors into the upper atmosphere for scientific purposes.

The first clearly defined scientific programme in space began with the International Geophysical Year, following the launch of the satellite Sputnik I. The origin of the IGY can be traced through the earlier international collaborations such as the Polar Years. As already mentioned, the Year is the expression of an international geophysical tradition, which ultimately stems from the shift in man's relationship with the earth, which began to make itself felt towards the close of the nineteenth century.

This new relationship between man and the earth was also making itself felt in other ways, and began to influence scientific thinking through these other channels.

The philosophy of man and space

Over 65 years before the first step into space was made,

Konstantin Tsiolkovsky began developing the scientific
principles of space flight. Many of his ideas on the
philosophy of man and space bear the mark of the developing
relationship between man and the earth as already
described. But some features of it, which were not of
importance in establishing the international geophysical
tradition, come to the forefront when man and space are
considered.

The relationship of man with the earth in this
particular context is as follows. The earth is man's
home, although he has always had to struggle to live on
it. Ever since prehistoric times man has survived in
this struggle by expanding outwards over the face of
the earth. Thus he escaped the unfavourable consequences
of his growing numbers, as well as of the exhaustion of
resources and the accidental or unthinking spoilation
of the environment. Such a course was followed, guided
increasingly by the earth sciences, until the last
decades of the nineteenth century. At that time, such
spreading became all but impossible because of the extent of human occupancy of the globe. Yet the old habits
continued, leading to crises, economic and social, within
and among nations.

Tsiolkovsky envisaged man reaching a new stage
of development in which these limitations would no

longer apply following the expansion of human civilization
beyond the earth. He said: "This planet is the cradle
of the human mind, but one cannot spend all one's life
in a cradle", and again, "Humanity will not always remain
on earth. In its desire to have more light and space
it will first penetrate beyond the atmosphere and then
it will conquer all the immense space within the solar
system" (KOSMODEMYANSKY, 1956, p. 95).

But we must recognize that even though man may
ultimately have his greatest fulfilment beyond the
earth, this planet will prove to be his test. The time
when he grows beyond the earth is still in the future,
despite the exploration and probing that is being done
today. In the meanwhile, terrestrial problems are multi-
plying at such a rate that they cannot be long left
without solution. It is becoming clear that man will
have to solve these problems long before the time comes
when he can expand his society into space and grow beyond
his terrestrial infancy. As far as space is concerned,
earth-directed space science will be for some time to
come more important to man than is distant space travel.

The earth has until now governed man's evolution.
Man has been fortunate that his development has been
spurred to its present level by a largely favourable
environment. He has, however, reached a state where

this development could even stop or go astray unless
he assumes the responsibilities attendant on maturity
and begins to develop himself and the earth so as to
direct the future course of events towards constructive
ends. A turning point has been reached; one direction
leads to a false turn in development, the other leads to
a yet unknown, but potentially higher future. It is
still in man's power to choose. In the practical sphere,
it means developing the whole earth so that its fruits
will serve all mankind, and that all of its regions, however
unfavourable they may now appear to the unimaginative, can
be transformed and developed to serve human culture.

      With these fundamental challenges so bound up with
scientific questions, the philosophy of science has once
again assumed importance in human affairs. Because earth
science is one of the principal sciences caught up in this
new relationship, this science will more and more find itself at the centre of ideas regarding man and his relationship
to the world. The scientific study of the earth and its
origin holds the answer to major philosophical questions.
Otto J. SCHMIDT, one of the great originators of creative
ideas regarding the evolution of the earth and the solar
system, lists (1958, p. 9) the areas in which earth science
contributes to philosophy. As regards this contribution he
says: "that the problem of the earth's genesis is of tremendous

importance is axiomatic. From the standpoint of...scientific philosophy, it is one of the three most important in natural history, the other two being the origin of life on earth and the origin of man."

Until man can answer these and related questions posed in our present age, man will be unable to provide a truly lasting rationale for his future on the earth; and even more important for current concerns is the fact that without this rationale, he will be unable to control or understand even the first steps he takes into space. Until he can do these things, he will be unable to enter the future that he could have in this new world open to him. In such a process he must change himself; indeed this change may be a necessary condition for being able to evolve into a new being beyond the earth. We now have no choice but to develop our planet in a new way, as "the earth our garden" (CROWTHER, 1955, pp. 93-133). We must make its fruits permanently renewable, and ensure that its hostile aspects and forces (natural disasters) are controlled or at least mitigated. In meeting such needs, the earth sciences become elevated in importance. In the course of attempts to solve these problems, man will face the greatest test in his history. Earth sciences are a critical part of the solution, not only in the practical sphere but also in the development of philosophy. More than ever, man requires a heightened awareness of the nature of the earth and space beyond it. Because of this dual need, practical and philosophical, earth sciences are growing in support and prestige, and deserve recognition far beyond the bounds of science. Awareness of the need for these sciences is spreading even though at this early stage of man's most crucial period on the earth it is difficult to view completely and clearly the sources of this new stage in the development of the earth sciences.

The earth's crust and its influence on science

Besides providing a theatre on the cosmic scale for experiments giving new insights into science, the scale of the earth (and of the processes which formed it) because of its influence on the materials composing the earth, influenced science in still other ways. Whenever science has been linked to a sufficient extent with problems related to the study or the exploitation of the earth's crust, it has been offered challenges which otherwise would not have been encountered. In meeting these challenges, science has often been stimulated to develop in new ways, with these developments leading to fundamental discoveries.

As an example, we may cite the development of
some of the early phases of chemistry. The scale of time
involved in the formation of rocks gave rise to crystal
forms with particular characteristics. The laws of
migration of the elements during the evolution of the
earth led to association of elements which helped shape
the course followed by the development of chemistry.
The chemical elements are distributed in a systematic
way across the face of the globe; they occur in
characteristic amounts, associations, and relative
proportions in different regions. Speaking broadly we
see differences in chemical composition among continental
shield areas, continental platform areas, subcontinental
areas, and oceanic areas. If we consider greater detail,
we see further subdivisions; for example, within a particular shield area, subdivisions are found which differ
one from the other in their chemical characteristics. The
earth may be viewed as a vast chemical system, throughout
which the chemical elements are distributed in an ordered
fashion. The migration of elements during the evolution of
the earth is controlled in part by properties such as atomic
weight, atomic or ionic radius, and ionic charge. These
properties also are a principal group among those determining
the position of the elements in the Periodic Table. In some
cases then we might expect the systematic distribution of

elements as seen in the Periodic Table to be reflected geographically. As an illustration of this, A.E. FERSMAN (1958, p.66) one of the pioneers of geochemistry, speaks of the distribution of the chemical elements in the Ural Mountains area as appearing before us "as an enormous Mendeleyev's periodic table spreading across the rocks". Furthermore, considering greater detail still, we see characteristic associations of elements occurring in various types of ore deposit. Since much of inorganic chemistry developed originally from problems arising from the analysis and the benefaction of ores and minerals, it is not surprising that many of the leading figures in the history of chemistry, including Dmitry Mendeleyev, were greatly interested in natural resources and in the chemistry of natural products. Mendeleyev, for example, had an extensive knowledge of petroleum and of industrial and metallic minerals, and took part in exploration for natural resources.

A second feature of the chemistry of the earth's crust, which influenced the development of chemistry, is the fact that most of the crust is crystalline in nature. The study of minerals and crystals is one of the oldest branches of earth science, and many other branches of science have stemmed from this source.

Having seen a few cases which point to a close

connection between the earth's crust and chemistry, let us look more closely at some of those aspects of the history of chemistry which were conditioned by that relationship.

In considering the history of a science, both the chain of technical and social influences, and the lines of influence from science itself must be considered. Bernal's (1965, p. 974) Table of the development of science shows clearly the interweaving of influences that occurred in the development of chemistry. Two technical chains, beginning in prehistoric times: one leading through the discovery and use of fire, pottery, glass, drugs and dyes; and the other beginning with stone implements, and going through the discovery and technology of metal and of mining and smelting finally joined to create the early beginnings of chemistry in the Early Iron Age (BERNAL, 1965, p. 172). Mining, metallurgy and chemistry continued in a close relationship until the Renaissance, and regarding that time it has been noted that "the greatest advances of Renaissance technology were in the closely linked fields of mining, metallurgy, and chemistry...the smelting of metals was the real school of chemistry". There was a rapid opening up of mines in this period, and in the course of learning to smelt new ores and process new metals, there arose a "general

theory of chemistry, involving oxidations and reductions...
at first mainly implicitly assaying, to find the yield
of an ore in precious metals...became the basis for
chemical experiment and chemical analysis". There was
of course an interest in a great number of other fields
which had their influence on chemistry: medicines, pottery,
distilling, wine making, brewing, and gunpowder to
mention a few. Technique and ideas in chemistry were
thus developed during this long period with a strong
connection with mining and metallurgy. There was not,
however, a science of chemistry in the modern sense, and
this did not come until the appropriate comprehensive
ideas had been developed, when the Industrial Revolution
was approaching and entering its first phases.

At the time of the Industrial Revolution, a strong
social and technological stimulus began to develop and
initiated the consolidation of a number of arts and sciences
into what was to become modern chemistry. "The interest
in chemistry was reflected in industry, and in turn, industry
supplied chemistry with new substances and new problems"
(BERNAL, 1965, p. 451). The influences for the development
of the new theories came mainly from physics, with the
pneumatic revolution (in which combustion was recognized as
a process of combination with oxygen) brought to the
decisive point by A. L. Lavoisier, who himself came to

chemistry from physics. The second great theoretical
development was the atomic theory of John Dalton, followed
by a third: the idea that electric charge could be
carried by atoms. This development soon led to the idea
that salts were formed by the mutual neutralization of
positive and negative charges. From this point,
J. Berzelius went on to develop the basis for modern
inorganic and mineral chemistry in the early decades of
the nineteenth century.

Another science, crystallography, brought into
its modern form by R. J. Haüy in 1800, fed an influence into
chemistry through the idea of isomorphism in which
"crystallography could become a useful adjunct to
chemistry" (BERNAL, 1965, p. 453). Other growing
industries of the Industrial Revolution, particularly
the textile industry, combined with a continuing interest
in medicines, and the growing interest in scientific
farming and fertilizers, led to the development of
organic chemistry. There was by this time a general
enough theory to unite all these diverse chemical fields
into a single science of chemistry.

One of the most fundamental developments in the
new science was the formulation of the Periodic Law, the
history of which has been traced by PARTINGTON (1964,
vol. 4, pp. 894-897). Partington describes its beginning

with Prout's hypothesis in 1815, continuing through the
work of a number of scientists including J.A.R. Newlands,
and Lothar Meyer. Most of this work was based on the
numerical analysis of atomic weights, and the theory
was not widely accepted because of doubts thrown on it by
the suggestion that a chance association of numbers, some
of which were not at that time well known, was giving
rise to the false appearance of order. General acceptance of
the Periodic Law might not have come for a considerable
time after it did if it were not for the establishment
of a more widely based foundation, by Dmitry Mendeleyev.
His arrangement of elements was more firmly based on
chemical properties and less on mere numerical analysis
of atomic weights than were those his predecessors.
The scientific advance represented by Mendeleyev's approach
is detailed by WEEKS (1968, p. 636). Mendeleyev himself
provides an illuminating commentary on the scientific
basis for his particular approach to the Periodic Law.
The extent of his departure from the merely numerical
approach is illustrated in his Principles of Chemistry
(1897, p.1), where he lists the relation of the atomic
weights of the elements as the last of the four properties
of the elements and their compounds which he judged as
the most important in establishing the Periodic Law.
He lists isomorphism, which he called "the analogy of

crystalline forms" as the first of the four properties,
remarking that isomorphism was "historically the first,
and an important and convincing, method for finding a
relationship between the compounds of two different
elements". He traces the origin of the idea of
isomorphism to the study of crystals remarking that this
property has "more than once been employed for discovering
the analogy of elements and their compounds". He then
pointed to the earth as the origin of the interest of
scientists in isomorphism, stating that this interest
arose initially because of "the property of solids of
occurring in regular crystalline forms...(and of)...
the occurrence of many substances in the earth's crust
in these forms...(thus long ago attracting)...the attention
of the naturalist to crystals...Thus was established the
idea of isomorphism as an analogy of forms by reason of
a resemblance of atomic composition, and by it was
explained the variability of the composition of a number
of minerals as isomorphous mixtures".

This recognition on the part of Mendeleyev, in
minerals as isomorphous mixtures went back far in his
scientific career. In fact, (PISARZHEVSKY, 1954, p. 4)
notes that the first chemical researches of Mendeleyev
were on minerals. His first paper, in 1855, was
entitled "The Analysis of Finnish Allanite and Pyroxene".

The phenomenon of isomorphism was prominent in the chemical problems related to these minerals, and he continued his researches on this problem, leading to his first major contribution to chemistry. Minerals give some of the best examples of isomorphism, and "the phenomenon of isomorphism which attracted the young scientific worker's attention clearly reveals the similarity in behaviour of atoms of different elements. Later Mendeleyev referred to isomorphism as one of the most important characteristics on the basis of which elements may be grouped into their natural order."

A good many further examples have been pointed out in which scientific ideas of the first rank arose in geology because of the time-scale involved in the development of the earth's crust. Two such ideas are the concepts of evolution and of geologic time (HAGNER, 1963, pp. 233-234). Both ideas can be traced far back into human history, but they first became powerful ideas in science when they were applied to the geologic record. This record contains the results of many natural experiments, where long stretches of time have made it possible for natural processes to work themselves out in a way which would not be possible in the smaller time-scale available in ordinary laboratory experiments or without the special nature of materials of the earth's crust.

In this way, many geological observations have made major contributions to general scientific ideas (McKELVEY, 1963, pp. 71-73). These contributions are jointly the result of the earth's position in the system of levels of scale, and of its particular composition and material state.

Chapter 2

THE SCIENTIFIC REVOLUTION (1450-1700)

2.1 The importance of science and history during the Scientific Revolution

We have seen briefly in the last chapter an outline of the development of earth science, and some of the principles governing the development of these sciences. One of the basic principles is that of the parallelism between the history of science on one hand and general and economic history on the other. The Scientific Revolution is a period which has been extensively researched in the History of Science, and is the first period in which modern earth science began to consolidate. It is therefore appropriate to begin with a fairly detailed account of some of the earth sciences and their place in the society of that time.

2.2 A transformation of science

During the two and one-half centuries following 1450, there occurred one of the most remarkable developments of science ever recorded. Science had, during the Middle Ages, developed in Europe at a very much slower

pace than it had previously done in the best periods
of the classical civilizations. In fact, much of the
science that was to be the background of the new development was transmitted to Europe from classical sources
and the Far East by the Islamic civilization, which
enjoyed a cultural flowering during the Middle Ages.
It should not be assumed, however, that the Middle
Ages in Europe had no influence whatever on science.
Although science was by and large dormant at that time
in Europe, the foundations for further advance were
being laid. Principles of scientific method were formulated and many technical developments took place (BERNAL,
1965, pp. 228-250). For a number of reasons great changes
began to occur, gathering momentum by the end of the
fourteenth century. These changes were the result of
political, economic and religious factors as they existed
in Europe at the time. They gave rise to a new spirit
of enterprise and to new intellectual horizons which
formed part of a great process of change which "inaugu-
rated a new order in economy and science" (BERNAL, 1965,
p. 251).

These changes are now seen to have been the first
stages of a great transformation of science, which has
come to be known as the Scientific Revolution. The
Scientific Revolution occurred in a number of phases.

It began with what is perhaps its definitive phase,
encompassing the Renaissance. During this period, science
challenged old authority and embarked upon a fresh
and inspired descriptive survey of the whole range of
human endeavour and experience. Most significant in
their effect upon science were the great navigations,
which led to the discovery of the Americas, the circum-
navigations of the globe, the descriptions of new lands,
and the many voyages in search of trade and colonies.
The impact of these developments was felt in many ways
in a variety of scientific and technical fields.
Significant in their influence were navigation and map
making, which extended challenging problems to the
sciences.

2.3 The great navigations and the earth sciences

Influence on science in general

The great process of change reached through all phases
of life, stimulating the development of a whole range
of scientific disciplines. Of particular interest in
tracing the origin of earth sciences is the outward
thrust in exploration and trade which occurred at the
time. This, a predominantly maritime expansion, was
perhaps the most characteristic feature of the times. Its

effects were seen not only in the practical spheres of
trade and commerce and the closely related sciences concerned with surveying the earth, but also in general ideas.
The view of what was possible extended its range many times
because of the physical breaking of geographical barriers
brought about by the maritime expansion. This overseas
movement began to gather momentum towards the end of the
fifteenth century. In the little more than three decades
from 1488 to 1522, the Cape of Good Hope had been rounded
for the first time since (as legend has it) Carthaginian
times; the Americas had been discovered; and the world had
been circumnavigated for the first time. These, the great
navigations, had a profound effect on science.

Francis Bacon, one of the "major figures...at the
turning point between medieval and modern science"
(BERNAL, 1965, p. 304) was acutely aware of the new
horizons in science revealed by the great navigations.
He remarked that "it would be disgraceful if, while the
regions of the material globe - that is of the earth,
of the sea, and of the stars - have been in our times
laid widely open and revealed, the intellectual globe
should remain shut up within the limits of old
discoveries" (quoted by CROWTHER, 1960a, p. 109). In
fact, the scientist and the maritime discoverer of
those times were often close in outlook. Francis

Bacon and Sir Walter Raleigh conversed at length on the
latter's expeditions. "The discovery of the New World
had had a profound influence on Bacon's thought. He
transferred the spirit of the exploration of the earth
to the exploration of the whole universe of nature.
The idea of comprehensive discovery was common to the
thought of both Raleigh and Bacon, though they saw it
in different perspectives, one as the medium of adventure
and profitable enterprise, and the other as the means
to the control of nature in the interests of man".
(CROWTHER, J.G., 1960a, p. 291).

We have seen, through their influence on a major
figure at the point of origin of modern science, the
stimulation given by the great navigations to a broad
range of sciences. The navigations continued in this
role throughout the Scientific Revolution; as evidence
of their long-lasting influence on science we find them
listed among the principal reasons for founding the Royal
Society of London in 1662, almost four decades after
Bacon's death (1626) (BERNAL, 1965, pp. 320-321).

Influence on the scientific study of the earth

The broadening and growth of ideas affected all scientific
development, including studies of the earth. The latter,
because of their nature, were among the sciences directly

stimulated by the overseas movement. In particular, the sciences of navigation and of map making had direct application to overseas expansion. These two disciplines, along with the sciences needed to advance them, rose to be among the principal sources of science for those times. They further influenced the direction, scale and speed of scientific development because they offered the prospect of almost immediate application of any technical developments made during the solution of these problems. The great developments of early modern astronomy were inspired by problems posed in response to the practical needs of this great age of navigation. Such a setting for the science of the age was to lead at a later stage to the establishment of the earliest branches of the earth sciences. The initial largely observational phases of these disciplines developed as part of the great wave of navigation and geographical exploration. Their theoretical foundations were subsequently laid once sufficient advance had occurred in sister sciences such as mechanics, astronomy, and cartography. These sister sciences were themselves stimulated into growth to a considerable extent by the same social developments which were giving rise to the earth-directed sciences. The central problems of these branches of earth science were solved before the end of the Scientific Revolution; simultaneously the other related sciences provided the techniques of experiment and mathematical analysis which assured that solid advance would be made in studies of the earth.

The situation in science generally was a culmination of the advances made throughout the Scientific Revolution in a period of remarkably swift advance during the last fifty years of the period. This short span marked the "definitive phase in the establishment of modern science" (BERNAL, 1965, p. 310).

The Scientific Revolution was the age of great and lasting innovations in science. This was the age of Mercator's maps, Copernicus' solar system, Kepler's planetary orbits, Galileo's telescopic observations, as well as dynamics, Boyle's gas laws, and Newton's theory of gravitation.

The branches of earth science related to the earth's shape and gravitation, as well as to geomagnetism, were established during the Scientific Revolution. A number of fundamental scientific problems in these areas were recognized, and methods of observation and theoretical analysis were established in forms that were to determine developments for some time afterwards. Several lines of development can be traced from that period to the present day. Other sciences of the earth such as mineralogy, and some parts of geology, meteorology, oceanography, vulcanology, and seismology all had, at least as sporadic appearances,

recognizable forerunners during (and in some cases before)
the Scientific Revolution. But their establishment into
sciences was to await a later period.

But in the beginnings that were made, the Scientific
Revolution, in a relatively short time of rapid growth, set
the problems and methods and lines of advance in earth science
which continued for a long time afterwards. The influence
of navigation, through the technical problems it raised, is
visible everywhere in the earth science of the Scientific
Revolution and the periods immediately following.

2.4 The earth sciences during the Scientific Revolution

The two branches of earth science: those dealing with
geomagnetism and with the earth's shape and gravity,
which were founded during the Scientific Revolution,
will be traced beginning with their forerunners and then
through the establishment of the basic sciences behind
them until the time at which they evolved into true
earth sciences.

The relationship between navigation and earth
science in the Scientific Revolution has been emphasized.
It should be made clear, however, that the development of
science seldom depends simply on one factor alone. The
thread of development of any branch of science follows a

complex and often seemingly erratic path. Sometimes the
thread is strong and continuous; at other times it
diminishes in size, and may even die out. If it dies out,
it may reappear unexpectedly at another time or place.
How the threads in earth science work their way through
time and space will be examined (in the next chapter),
once we are in a position to sum up the position of earth
science at the end of the Scientific Revolution, a fruitful and critical period of history. The continuation or
reappearance of these thread of development is dependent
on a great many factors. Sometimes such development may
be seen to be linked with the efforts of a scientific
society, or with the researches, privately financed, of
a scientist of independent means. Alternatively it may
have been kept alive by a large state-subsidized insti-
tution, or by a university. Even more remarkable is the
fact that this track of development, erratic though it
might appear, lies close to the average track of
development of society as recorded in general history.
An examination of the relationship between these tracks
of development allows us to distinguish those historical
circumstances which favour the development of science
from those which do not. There must be, if development
is to occur, the proper combination of material incentive
and support as well as intellectual stimulation. These

factors are a part of the general state and spirit of an age. If there is an insufficient development of the required factors, then we may see some disciplines, once flourishing, fall into disuse because of the lack of suitable applications and the consequent lack of opportunities for the scientific results to be consolidated into the broader sphere of technology or other applications. The direction taken, or the rate of development, may be further influenced by certain conditions within or immediately surrounding the science. For rapid development to occur, sufficient and suitable previous growth of the science itself(or its predecessors)as well as of supporting sciences must have occurred. There must also be a body of scientific workers, either already suitably trained or experienced, or potentially available, to carry out the developments.

Several branches of earth science first developed during the Scientific Revolution, and we will begin by examining them. Their development will be traced first by tracing the stories of their growth and then following in later chapters with a detailed analysis using growth curves correlated with the general development of science at the time.

2.5 Geomagnetism

History of the magnetic compass

Geomagnetism in modern times developed as an outgrowth
of interest in the magnetic compass, and was perhaps of
all the sciences the most intimately connected at its
inception with the needs and spirit of maritime enterprise at the time of the Scientific Revolution.

The compass is, by its very nature, intimately
connected with the earth. The present-day observer, when
considering the history of the compass, should take into
account the very high degree of interest aroused at the
time of its introduction by the remarkable directional
property of the instrument. This interest ultimately
led to new ideas about the nature of the earth.

The magnetic compass was invented in China perhaps
as early as 100 B.C. (MASON, 1953, p. 54). The general
north-south alignment of a magnetized needle was recognized
from the first. Further refinements in the understanding
of the compass came at a later date. The declination, or
deviation of compass direction from true north, a quantity
which varies with position on the earth's surface, was
recognized at least as early as 720 A.D., when the first
recorded value of declination was made (NEEDHAM, 1965a, p.
301 and our Section 1.2). An interesting geophysical

test of this and subsequent recorded determinations of declination in China from the eighth until the nineteenth centuries has been published by SMITH and NEEDHAM (1967), concluding that the early observers in China had perceived and correctly recorded an important geophysical phenomenon long before anyone else was even aware of it.

The compass was introduced into Europe in (or somewhat before) the first part of the twelfth century, as far as is known from the earliest records (BERNAL, 1965, p. 235). The mode of travel of the invention from China to Europe is unknown, but apparently the knowledge of declination was not transmitted with it. The magnetic compass received its first mention in Europe in a scientific treatise just prior to the thirteenth century, by Alexander Neckam in 1190.

The directive property of a magnetized needle

Initially, the directivity was explained as being due to the influence of the heavens. The northern pole of the Ptolemaic system of spheres, and a number of polar stars were suggested at various times as the point toward which the needle turns. Such views were put forward until at least the mid-sixteenth century, and were dominant as long as Ptolemaic cosmology was dominant. In this view, the earth was seen as a passive body centred beneath a

system of spheres mysteriously rotating above it. For
example, Peter de Mericourt (Petrus Peregrinus) believed
that the celestial pole is the point to which the needle
is drawn. De Mericourt was a pioneer in magnetism and an
experimentist. He discovered magnetic meridians on a
magnetized sphere, developed the concept of poles and
knew that unlike poles attract and like poles repel
(MASON, 1953, p. 88). He developed these ideas in his
book Epistola de Magnete (1269), which has been described
as "the first original scientific work of western
Christiandom" (BERNAL, 1965, p. 235). But in his time
the end of the Ptolmaic system was not in sight and
even such a talented scientist could not extend his ideas
beyond it.

Some writers revived classical legends of lodestone
mountains, using these an an explanation of the directivity
of the compass. GILBERT (1600, p. 9) remarks on the pro-
ponents of these ideas, who in his view erroneously imagine
"the existence of hyperborean magnetic mountains, attracting objects of magnetic iron... imagining to themselves
the existence of magnetic poles and mighty cliffs." These
views were seen by Gilbert to have no scientific basis and
had in fact been dismissed on the basis of logical arguments by de Mericourt in 1269. It has been suggested
(SMITH, 1968, p. 505) that the revival of the legends of

lodestone mountains served a useful purpose in the study of the earth, directing attention from the heavens to the earth as the seat of magnetic influence. However, the Copernican revolution of the sixteenth century probably was much more powerful, acting as the primary cause of the evolution of views on magnetic directivity. These ideas underwent a great change from those of de Mericourt (published in 1269), who saw the magnetic influence lying in the heavens, to those of Gilbert (published in 1600), who seated it in the earth. In the Copernican revolution astronomy was freed of the old contradictory picture of the earth as being at the centre of celestial motion yet not serving as the seat or theatre for the operation of celestial forces. GILBERT (1600, p. 322) represented the new view, and said of the old: "...as regards this primum mobile with its contrary and most rapid career, -where are the bodies...that propel it? ...and what mad force lies beyond the primum mobile? - for the agent force abides in bodies themselves, not in space, not in the interspaces." Aided by this outlook Gilbert was able to recognize in the results of his experiments, in which he explored the magnetic field at the surface of spheres ("terrellae", or "little earths" as he called them) cut from lodestone, that the earth is a great magnet. He saw that the directivity of the compass could be only terrestrial in origin, and that the influence responsible for the effect must be global in nature.

During the century preceding Gilbert's de Magnete, new developments in the practical application of the compass were taking place. The compass had been used on land and at sea since its introduction into Europe, and beginning with the great voyages, had been used in widely separated parts of the earth. These applications led ultimately to the widespread recognition of the declination and inclination of a magnetized needle and, in the century following de Magnete, to the development of terrestrial magnetism into an earth science.

Magnetic declination and inclination

The use of the compass in Europe and on voyages to distant parts of the earth led very quickly to a great interest in magnetic declination and to its incorporation into European science. The first recorded declination of the compass carried out by a European observer was made in Rome in 1510 (Table 2.1). The existence of the declination was, however, known before this date. Sundials made in Nuremberg from about 1450 have declination markings engraved on them (to allow proper alignment of the dials by magnetic compass) (MITCHELL, 1937, p. 271; 1939, p. 77). The compass developed rapidly during the period of the great

Table 2.1

Early declinations in Europe (from HELLMANN, 1899)

Year	Place	Magnetic Declination	Observer or Authority
1510±	Rome	6° E	Georg Hartman .
1518±	Bay of Guinea	11 1/4° E	Piero di Giovanni d'Antonio di Dino.
1520±	Vienna	4° E	Johann Georg Tannstetter.
1524±	Landshut (Bav.)	10° E	Petrus Apianus (Bienewitz
1534	Dieppe	10° E	Francois or Crignon.
1537	Florence	9° E	Mauro (Sphera volgare nov mente tradotta. Venetia 1537. 4°. fol. 53ª).
1538	Lisbon	7 1/2° E	Pedro Nunes or Joao de Castro.
1539	Dantzic	13° E	Georg Joachim Rheticus.
1541	Paris	7° E	Hieronymus Bellarmatus.
1544±	Nuremberg	10° E	Georg Hartman .
1546±	Is. of Walcheren	9° E	Gerhard Mercator.

voyages, and by 1525 compasses of sufficient reliability
to measure declination at sea were being made. This last
development made it possible to increase significantly the
rate of coverage of the globe with determinations of declination. The first authenticated map of declination was
prepared by Alonzo de Santa Cruz in 1536 (MITCHELL, 1937,
p. 270). By this time it was becoming recognized that

declination varies with location, forming a definite pattern
of values over the earth's surface. This feature of declination raises the possibility that geographical points
might be fixed from magnetic measurements.

The determination of geographical coordinates could at
the time be done only by astronomical methods, which were
difficult to carry out and often inaccurate. The possible
application of magnetic measurements to this difficult
problem was therefore a matter of great interest. We shall
see in later sections that the method never proved to be
practical, although the possibility that it could be
applied spurred geomagnetic measurement in the sixteenth
and seventeenth centuries. The most direct and lasting
contribution of a worldwide knowledge of declination has
been in making it possible to steer a true course guided by
the magnetic compass. This application also spurred geomagnetism greatly. As a result of this growth of interest,
charts of declination began to be published at frequent
intervals.

One notable sea voyage was that of Joao de Castro
who commanded one of eleven ships that sailed from Portugal
to the East Indies in 1538. He was commissioned to test
thoroughly the performance of the new marine magnetic
compasses and to test the magnetic method of determination
of longitude. He recorded a series of 43 determinations

of declination throughout the voyage, between the years
1538-41 (CHAPMAN and BARTELS, 1940, p. 909).

In this way observations of declination were,
during the remainder of the century and through the
following century, carried over the globe by mariners,
explorers and scientists. This expansion, ultimately
motivated outside science, led to the observation of
magnetic phenomena over a large portion of the earth's
surface. These observations, once they were sufficiently
extended, began to reveal phenomena on the scale of the
earth.

As a result, magnetic studies of the earth began
to show the features of a separate science; geomagnetism,
the first of the earth sciences, began to emerge. Notable
among the beginnings of earth science were the theories
of the origin of the terrestrial magnetic field developed
by Edmund Halley and by Robert Hooke. Let us look first
at the earliest example of a comprehensive geophysical
survey on the earth-scale, conducted by Edmund Halley
in 1698-1700.

Edmund Halley's chart of declination
The world-wide distribution of declination was of
particular concern to a maritime nation like England.
The request for assistance in procuring a vessel to study

magnetic variation was made in 1693 to the Royal Society by Benjamin Middleton, with Edmund Halley shortly after requesting to join the enterprise. A craft was built and fitted out under the auspices of the Admiralty. It was originally understood that the ship, the Paramour, would sail at the expense of the Royal Society. However, by the time that the expedition set off, the enterprise had come under the patronage of the King, and for reasons not yet fully understood, Middleton left the enterprise and Halley assumed command (ARMITAGE, 1966, p. 139).

The expedition sailed for two years (from 1698 to 1700) in the North and South Atlantic, covering these areas with a network of sailing tracks along which measurements of declination were made. Halley published the results a year or two after his return as an isogonic chart of the Atlantic Ocean (ARMITAGE, 1966, p. 147). Further, Halley had been for some time before his voyage compiling reported declination determinations from all over the world. He added these to his own observations and published an iso-gonic chart of declination for the world, about the same time as the Atlantic chart.

The influence of instrument making on the development of geomagnetism

The general expansion of science and technology during the

Scientific Revolution led to a radical change in the
scale and scope of the instrument making trade. Its
growth is outlined by PRICE (1957, pp. 582-647). He
notes that at the beginning of the sixteenth century
there were "two distinct types of instrument-makers.
On the one hand there were scientists... whose special
interest was in the design and actual making of instruments. On the other hand there were whole dynasties of
craftsmen..." (PRICE, 1957, p. 621). But by 1650, "the
full impact of the scientific revolution manifested itself
in the instrument-making trade by great changes in its
scale and scope... The greatest effect of the scientific
revolution was wrought by new instruments and discoveries
...(particularly)... optical instruments - the telescope
and microscope... (and)... new practitioners' instruments
for surveying, navigation and gunnery" (PRICE, 1957,
pp. 629-630).

The discovery of magnetic inclination

This phenomenon, in which a magnetized needle free to
turn in all directions comes to rest not only pointing
to magnetic north but along a line dipping below the
horizon, is due to the fact that the needle seeks
alignment with the curved lines of force of the geomagnetic
field. This phenomenon is one which would have become

easily discoverable even by accident any time after the
rapid improvement of compasses that began early in the
sixteenth century. A pivoted needle type of compass was
described as early as 1190 by Alexander Neckam. The balance
of such a needle is affected by the inclination of the field.
However, the effect might easily have been missed in the
earlier cruder instruments. It was in fact in the sixteenth
century, after the instrument makers began to contribute
their improvements, that the discovery was made. Two widely
separated observers, apparently unknown to each other, made
this discovery. George Hartmann, vicar of St. Sebaldus,
Nuremberg, made the earliest documented observation of magnetic inclination, in 1544 (HARRADON, 1943, p. 128). This
observation was followed by an independent one by Robert
Norman, a retired mariner who became an instrument maker,
in 1576 in London. He performed the first recorded accurate
measurement of inclination, and published a description of
the phenomenon in his book The Newe Attractive in 1581.
Norman's contribution influenced geomagnetism rapidly because it occurred in the midst of maritime expansion in a
country immersed at the time in overseas expansion. Thus it
led to immediate application and thereby became part of geomagnetic theory on the planetary scale, for example in Gilbert's
synthesis of magnetic knowledge. Also, measurements of
inclination were made in widely separated regions of the earth.

Interest in geomagnetism and in its possible contribution to navigation was high. The potential applications of geomagnetism were expressed by William Gilbert as follows.

Gilbert believed that maps of magnetic declination and inclination could be used to provide the position of a ship at sea. This idea, as the authors of one of the standard books on geomagnetism state, "is theoretically sound but its practical application is hindered by the inaccuracy of the measurements in ships, due to their motion (and their iron)" (CHAPMAN and BARTELS, 1940, p. 913). In discussing the question "whether the terrestrial longitude can be found from the variation", Gilbert says "Grateful would be this work to seamen; and would bring the greatest advance to Geography...(the method)...is of great moment, if only proper instruments are in readiness, by which the magnetick deviation can be ascertained with certainty at sea". Gilbert further suggests a method of employing measurements of magnetic inclination "To ascertain the elevation of the pole or the latitude of a place anywhere in the world, without the help of the celestial bodies, sun, planets, or fixed stars, in fog and darkness" (GILBERT, 1600, pp. 251-253).

Gilbert's prediction from his experiments with his terrellae, that magnetic inclination should vary with latitude in a regular manner, was confirmed by Henry Hudson on his voyage to North America in 1609.

Secular variation of the magnetic elements

By the beginning of the seventeenth century, over a hundred
years of observations of magnetic declination had accumu-
lated. This period of time lies within the time scale
of many of the planetary magnetic processes. In 1635
Henry Gellibrand, Professor of Astronomy in Gresham College,
London, by comparison with earlier measurements, discovered
that there is a secular variation of the magnetic declination. The magnetic declination had varied steadily westward
in London since 1580 when it was observed to be 11 1/4°E by
William Borough, and since 1622 when it was measured as 6°E
by Edmund Gunter. The value obtained by Henry Gellibrand
in 1635 was 4°E. EDMUND HALLEY (1692) in noting the
above values also lists subsequent values as 0° in 1657,
2 1/2°W in 1672 and 6°W in 1692, and remarks that "in
112 years the direction of the needle has changed no
less than 17 degrees". He cites further instances from
various parts of the earth which show "great changes in the
needles direction within the last century of years, not
only at London...but almost all over the Globe of Earth".

An earth science is born

Halley had earlier compiled values of all determinations of
declination known to him, having published a Table of Variations (HALLEY, 1683) showing changes of declination in

position on earth surface and in time. His observations
spanned the earth in space and time on the planetary scale.
He was therefore able to note phenomena on a sufficiently
high level of scale (Section 1.5) that earth science in the
true sense became possible. His work followed from that of
William Gilbert in viewing the earth as a great magnet, with
its magnetic force concentrated at certain points of
attraction, or poles. In order to explain the observations
he had gathered, Halley concluded that "the whole globe of
the Earth is one great magnet, having Four magnetical poles,
or points of attraction, near each pole of the Equator (are)
Two; and that, in those parts of the World which lye near
adjacent to any one of these magnetical poles, the needle is
governed thereby, the nearest pole being always predominant
over the more remote." (HALLEY, 1683,p. 215, and our
Figure 2.1). He remarked that it was, in his day, impossible
to calculate exactly the effect of the poles at any
locality since it "remains undetermined in what proportions
the attractive power decreases, as you remove from the
Pole of a magnet". This latter point had to await subsequent
developments: in particular a sufficiently sensitive device
for the measurement of the attractive forces exhibited by
magnets was required for a definite answer. Halley had
experimented with such measurements, using a needle device,
but without final or certain results (ARMITAGE, 1966, p. 75).

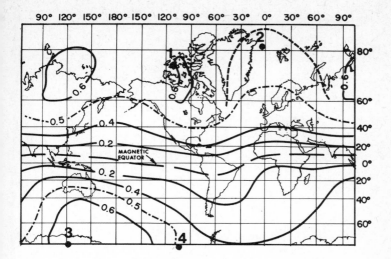

Fig. 2.1   Vertical component of geomagnetic field,
            with Edmund Halley's four poles superimposed.
            Maps of field from VESTINE, E. H., 1960.
            The survey of the geomagnetic field in
            space. Transactions, American Geophysical
            Union, 41(1), by permission of the
            American Geophysical Union.

John Michell in 1750 and, later, Charles Augustin Coulomb

in 1777, published results of the first successful measure-

ments of the attraction of a magnet, using a torsion balance.

This instrument was the one which made sufficiently precise

measurements possible. No less important was the advance

of ideas, allowing a clear view of the nature of magnetism.

Michell began the attack on earlier ideas of a vortex

theory of magnetism. This attack was completed by Coulomb, "wedding the physics of electricity and magnetism to the Newtonian idea of action at a distance" (GILLMORE, 1971, p. 194). Having possession of a suitable theory, Michell and later Coulomb were able to propose the inverse square law of magnetic attraction.

Halley's compilation of declinations indicated "yet a further difficultie" to him. It appeared "from the foregoing table...that all the magnetical Poles had a motion westward: but if it be so, tis evident that it is not a rotation about the axis of the Earth; for then the variations would continue the same in the same parallel of latitude...but the contrary is found by experience...whether these magnetical poles move altogether with one motion, or with several...are secrets yet utterly unknown to mankind" (HALLEY, 1683, p. 220).

Almost a decade later Halley put forward an explanation of these changes in declination, and of the difficulty "that no magnet I had ever seen or heard of, had more than two opposite Poles; whereas the Earth has visibly four, and perhaps more" (HALLEY, 1692, p. 564). As regards the time variations, Halley suggested that "the External Parts of the Globe may well be reckoned as the Shell, and the Internal as a Nucleus or inner Globe included within ours, with a fluid medium

between...For if this exterior Shell of Earth be a Magnet
having its Poles at a distance from the Poles of Diurnal
Rotation; and if the Internal Nucleus be likewise a
magnet having its poles in two other places distant
also from the Axis" (HALLEY, 1692, p. 568). Small
differences in angular velocity between the outer shell
and inner nucleus would, Halley thought, explain the
westward drift with time of some features of the geomagnetic field.

When viewed from the present-day position of earth
science, many features of these early ideas are found in
modern theories of the secular variation of the geomagnetic
field (Figure 2.2). It is not surprising that ideas at
this level could occur early in the history of geomagnetism
because the secular variations involved are large enough
to be measured by the simplest geomagnetic instruments.
The rate of overseas voyages was sufficient to allow a
large enough spread of observations over the globe in a
relatively short time so that the global spread of geomagnetic quantities could be determined for a correspondingly
short time period. By comparison with another time period
secular variation could be determined. Thus the observation
of geomagnetic phenomena on a planetary scale in both time
and space was possible, leading to the creation of an earth
science.

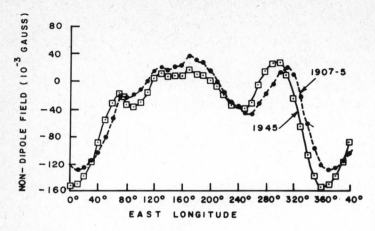

Fig. 2.2   Westward drift of the geomagnetic field.
             From BULLARD, E. C., his paper in the
             Philosophical Transactions of the Royal
             Society of London, 243: 79, by permission
             of the Royal Society and the author.

Thus in the short period of time from 1683 until
1700 Edmund Halley had proposed a theory for the origin of
the geomagnetic field and carried out a widespread programme
of measurements which marked his efforts as a part of
earth science in the full sense. We may take this interval
of seventeen years as the founding period for the first
branch of earth science fully in its modern sense.

Subsequent developments in the eighteenth century
Soon after Halley's map of declination the first map of

inclination appeared. It was made by William Whiston in
1721. The map covered southeast England and was made by
him in an attempt to solve the ever-present problem, as
yet unsolved in the first quarter of the eighteenth century,
of determining longitude at sea. This problem was the
primary one which motivated his work on inclination. He
referred to Halley's "Map of the Variation" and to how
it had "already given some Help for the Discovery of the
Longitude, at least near the Cape of Good Hope".
He remarked that the rapidity of change in declinations had
cut short the usefulness of the method in this one locality
where circumstances had once favoured its application.
Noting the fact that inclination changes more slowly with
time than does declination, he "conceived great hopes, that
this way of application of the Power before us might very
probably discover the Longitude" (WHISTON, quoted by HELLMANN,
1895, p. 12).

It is interesting to note that declination was in
fact used for a time after the publication of Halley's map
of declination (1701) for determining longitude. Only
declinations in areas where contour lines are directed north-
south proved to be suitable for this purpose. At the time
of Halley's map such an area was found off the Cape of Good
Hope. Subsequent changes in declination rendered the method
less useful and it fell into disuse. Thus a search for alter-

native methods, as exemplified in Whiston's work on inclination, was accelerated. Magnetic methods did not in the end prove to be sufficiently accurate measures of longitude. But the attempt to apply them for this purpose stimulated their development, thereby contributing to earth science. The development of geomagnetism was important to the science of the earth because of its many applications beyond the longitude problem.

Whiston's pioneering map was followed by a world chart of inclination prepared by Johann Carl Wilcke at Stockholm in 1768. Maps of inclination proved to be of limited usefulness for their original purpose of aiding determinations of longitude. However, they were of importance in earth science in extending understanding of the geomagnetic field. Wilcke's chart marks the end of the first phase of geomagnetism, begun in the Scientific Revolution, in which quantities measurable by simple needle instruments were mapped, and explained at least partially in terms of their distribution and secular variation, using poles and pole theory and simple models of the earth's interior.

Soon after the Scientific Revolution new phenomena were discovered. John Graham in London had found in 1722, after a series of careful observations, that besides the slow variation measured in terms of decades or centuries,

the geomagnetic field was subject to rapid variations
measured in hours or days. Examples of this type of variation
had been seen by earlier observers such as Edmund Halley
who had noted transient variations of the magnetic needle
accompanying auroral displays and following sunspot concen-
trations. A full explanation of these phenomena was not
forthcoming, however. As we now know, a knowledge of
electromagnetism (not then available) is required for even
a partial explanation. Furthermore, the measuring instruments of a new stage in electricity and magnetism (also
not available at that time) were required for progress in
measurement beyond simple dip and declination. Thus the
first phase of geomagnetism ended in the late eighteenth
century and a new initiative did not occur until the next
century when the required technical developments had been
made, following a new stage in social development, the
Industrial Revolution.

2.6   The earth's shape and gravity

Navigation on long voyages involving the plotting of
sailing tracks and the fixing of positions is dependent
on a knowledge of the shape of the earth and on the ability
to compare times observed at these positions with those
at other points on the earth's surface. The shape of the

earth and timekeeping are in turn closely connected with
the earth's rotation and gravitation. At the time of the
great navigations the development of methods of navigation
adequate for long sea voyages was a matter of the utmost
importance. Thus the sciences involved might be expected
to have been elevated to a position of first-rate importance
at that time, provided that the general level of science
and technology was sufficiently high to sustain a marked
and widespread development. Let us then examine the
history of some of the sciences related to navigation to
see if a significant acceleration of their growth did in
fact occur.

The earth's shape

The earth was viewed as a sphere by the ancient Greeks,
from the fifth century B.C. Greek map makers employed
projections using lines of latitude and longitude as early
as 300 B.C. The Alexandrian scientist, Eratosthenes, used
an astronomical-geodetic method in the following century
to measure the earth's radius. Using this determination
and a projection from a sphere to a plane, he constructed
a map of the known world, his map extending from latitude
10° to latitude 60° and between longitudes 10°W and 80°E.
The spherical form of the earth continued to be accepted
by the educated public from then onwards (SINGER, 1957,
pp. 504-507 and our Table 2.2).

TABLE 2.2

List of some determinations of the radius and arc
length per degree of latitude in the era of spherical geodesy.

Eratosthenes	-2nd century	Egypt
I-Hsing	+8th century	China
Al-Farghani and other observers under Caliph al-Ma'mun	+9th century	Middle East
Fernel	+ 1529	Europe
Snell	+ 1615	"
Norwood	+ 1635	"
Picard	+ 1669	"
Dominique & Jacques Cassini	+ 1684-1718	"
Regis and Jartoux	+ 1710	Manchuria

The last three items in that Table served to close
one era in geodesy and open another. Between 1650 and 1670
the rapid advance of instrument design carried astronomy
and surveying to a new level of accuracy (OLMSTED, 1942,
p. 123). This particular feature of the Scientific Revo-

lution has been described in some detail by PRICE (1957, pp. 629-630), and its effect on the development of geomagnetism has already been considered in the present book (Sec. 2.5). This technical development advanced observation in geomagnetism and geodesy to a markedly higher position on the scale of sophistication (Sec. 1.5). Such an advance increases the possibility that a science will lead to fundamental discoveries. Up to the time of this development all measurements of arc length assumed a perfectly spherical earth or avoided explicit mention of the question of shape. This period has been referred to as the <u>era of spherical</u> geodesy (LENZEN and MULTHAUF 1965).

A new era made possible by the advances in instrument making began in 1669 when Jean Picard, one of the pioneers of the new methods, carried out a survey of an arc of the meridian (for which he had been commissioned by the Academy of Sciences of Paris). In his survey he layed out a series of triangles from which he found the north-south distance between two parallels $1°22'55''$ apart, stretching roughly from Paris to Amiens. He employed the newly developed instruments which allowed him to determine the earth's radius and the arc length per degree with an accuracy far beyond anything achieved before. On completion of this measurement Picard recommended an extension of his arc through the whole of France. His recommendation was subsequently adopted and the work went ahead in 1684, continuing until its completion in 1718.

At the same time, scientific interest in the shape of the earth was growing and the idea that the earth's shape might deviate from a perfectly spherical form (not a new idea at all) received much scientific support. Isaac Newton had remarked in 1680 that the earth is either "spherical or not much oval" (CAJORI, 1928, p. 172). Astronomical observation was soon to lend further support to the idea that the earth may not be a perfect sphere. In 1691 Dominique Cassini remarked that Jupiter, from telescopic observation, appears to be flattened at the poles (GRANT, 1852, p. 245). Since Jupiter was known to rotate on its axis, similar flattening of the earth (also a rotating body) was strongly suggested by the observed form of Jupiter. Isaac Newton cited Jupiter's shape as an important indication that the earth might reasonably be regarded as similarly flattened (in Prop. XIX, Problem III of his Principia). Today we know that the earth is flattened at the poles and that as a consequence of the flattening the arc length per degree increases from equator to pole (Fig. 2.3).

Picard's proposal for extension of his arc was finally adopted and was carried out by a number of scientists including Dominique Cassini and his son Jacques during the period 1684-1718. A southern extension of the original arc

124

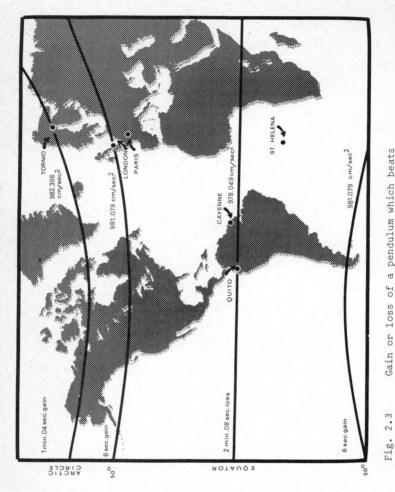

Fig. 2.3   Gain or loss of a pendulum which beats
seconds in Paris, with latitude. Also
shown is the latitudinal variation of
gravitational acceleration at sea level

carried it to the Pyrenees and the northern extension reached Dunkirk. The results indicated that the arc length on the northern part of the traverse was smaller than that on the southern part by a fraction of a percent. Today we know that it should have been a fraction of a percent larger. At that time there were no other known measurements carried out with sufficient scale and detail to afford a reliable comparison. The Chinese arc, surveyed in 723-726, approached the French arc in this regard, but it was unknown in Europe at the time. The Chinese results implied sphericity and might have, if known, led to serious reconsideration of the French arc. The results were taken at their face value, especially by Jacques Cassini (who became director of the Paris Observatory about 1710). The implications of these measurements were seen at the time to be that the polar diameter of the earth is longer than the equatorial diameter. This interpretation was in opposition to Newton's con- clusion in his Principia (1687) that the earth is flattened at the poles. The fact that extreme hostility existed between French and English scientists at the time no doubt reinforced the appeal in France of an equatorially flattened earth since the latter, thought to be based on experiment, ran counter to the ideas of Newton. The extent of this hostility is illustrated in the life of Voltaire. He had been unjustly forced into exile in England (as the

result of an incident not connected with science), and
returned as an embittered enemy of all arbitrariness. He
chose to use among other things the Newtonian view to
ridicule ideas in France after his return in 1729.
This view would not have been an effective weapon for
Voltaire if serious scientific differences had not existed.

In spite of the prevalence in France at the time
of support for Cassini's view as opposed to that of
Newton, the matter was not left to stand or fall sólely
on the French arc. The need for accurate world maps
felt by a maritime nation such as France, called for more
decisive measurements of the earth's shape. The French
arc after all spanned too short a spread of latitude
to settle the question decisively because the changes
in arc length per degree over the distance involved
are small; smaller in fact than errors that can easily
enter, unsuspected, into the methods employed. Such
errors in fact had entered into the determinations of the
French arc as was found later in remeasurements of the
arc conducted some seventy years after its completion
(TODHUNTER, 1873, p. 56, pp. 126-9). In 1733, the measurement of an arc of the meridian near the equator was
proposed to the Paris Academy (by Godin) as a comparison
with the French arc, because in this comparison the large
difference that would result in arc length would exceed

all reasonable errors in observation. In this application
the advances in techniques of measurement previously
referred to led to a new era of geodesy and of understanding
of the earth, at a higher degree on the scale of scientific
effectiveness. A relatively direct measurement of
the shape of the earth resulted from the proposed operations,
and they confirmed Newton's theoretical analysis (which
had been supported by experiment through his interpretation
of pendulum observations). By this chain of developments,
a new era in geodesy and in the understanding of the
earth, the ellipsoidal era of geodesy (LENZEN and MULTHAUF,
1965), was ushered in.

In the meantime, further proof of the ellipticity
of the earth had been obtained on the Manchurian plains
by Régis and Jartoux in 1710 (NEEDHAM, 1965a, p. 54).

The suggested expedition gained Royal assent and
left in 1735. Principal members of the party were Pierre
Bouguer, La Condamine, and Godin; the site of operations
chosen was equatorial South America (then a part of Peru,
now Ecuador), and two Spanish naval officers, Juan and
Don Antonio de Ulloa, accompanied them.

After this party, bound for the equator, left France,
it was decided to send a similar party to polar regions
to increase even further the separation in latitude of
the regions in which observations would be made. This

party was led by Maupertuis and included three other
Academicians: Alexis Claude Clairaut, Camus and Le Monnier.
They were accompanied also by l'Abbe Outhier, and Anders
Celsius, Professor of Astronomy at Uppsala. They left
Paris in 1736 for Lapland, where the operations, with
the aim of measuring the length of an arc of the meridian
at the Arctic circle, were to take place.

The arctic party, when it left France, was firm in
the conviction that Newton's views on polar flattening
were in error, and supported the stand, held strongly by
Jacques Cassini, that on the contrary the earth is elong-
ated at the poles. The group was first to complete its
operations (in January, 1737) and at first was astonished
by the results. The length of a degree of the meridian
proved to be larger, by a significant amount, than that
in France. As remarked by TODHUNTER (1873, p. 98) "The
party then went to... (its base)... and remained shut up
in their chambers in a kind of inaction until March.
The difference between their results and those suggested
by the Cassinian theory was so great that it astonished
them; and although they considered their operations to be
incontestable, yet they resolved to execute some rigorous
verifications". These verifications were carried out in
March and April, strongly confirming the original result,
and Newtonian theory. The geodetic operations were accom-

panied by pendulum experiments. Such experiments give
independent information on the earth's shape and were therefore an important part of the expedition's work. These too
confirmed the Newtonian theory; a pendulum regulated at Paris
to beat seconds, gained 59 oscillations in 24 hours as predicted
by that theory. The pendulum would have lost time if the
Cassinian theory were true (Fig. 2.3). Maupertuis became
a convinced Newtonian, as did other members of the expedition.
Clairaut later became a principal developer of the Newtonian
theory. It is significant that at the time when this
thread in French science (gravitation and figure of the
Earth) was becoming stronger, the same field were weakening
in Britain where Newtonian science had begun. The forefront
of development shifted to France, beginning with Clairaut,
where it continued through D'Alembert, Lagrange and Laplace,
culminating in the latter's Mechanique Celeste, in which
the spirit of dynamical laws as applied to the universe
were explored to their ultimate conclusion.

Voltaire, who had lived in England for a period
prior to 1729 and was deeply influenced by Newton and
Locke, was among the few Newtonians in France prior to
the Arctic Expedition. On the return of the latter,
Voltaire congratulated Maupertuis on having "flattened
the poles and the Cassinis" (TODHUNTER, 1873, p. 100).

The operations of the equatorial party, conducted

in Peru, (the region surveyed is now a part of Ecuador)
proved to be even more difficult than those of the northern
party. This circumstance, coupled with the remote and
difficult terrain in the Andes at the site of operations,
delayed their return until 1744. Their result, that the
length of a degree of the meridian is smaller than that
in France, was a further confirmation of the Newtonian
theory. In the course of operations, Pierre Bouguer
(in collaboration with La Condamine) carried out a series
of pendulum observations which led to a new line of
development in the study of the earth, to be treated
later (Chapter 3). Let us first consider the development
of the pendulum, an instrument which has played an important
part in the sciences of the earth.

The pendulum

This instrument played a large part in the development
of ideas on the earth's shape and gravity. Let us look
briefly at the information it can supply. The period,
or time for one oscillation, of a pendulum in its simplest
form varies as the following quantities:

(1) the square root of its length; and

(2) the inverse of the gravitational acceleration.
The relationship (1) above was discovered by Galileo
Galilei and published by him in his Dialogues concerning

two new sciences, completed in 1636 (CREW and DE SALVIO (trans.), 1914, pp. 94-8). The second relationship was discovered by Christiaan Huygens in 1659 (SOCIÉTÉ HOLLANDAISE, 1932a, p. 246).

Gravitational acceleration varies from the equator (where it is least) to the poles (where it is greatest) as shown in Figure 2.3. Thus a pendulum of a given length will beat more slowly at the equator than at the poles, with the rate varying smoothly between these extremes. For this reason, a pendulum clock regulated to run correctly at a high latitude will lose time if taken to a low latitude; this difference can be as large as a few minutes per day (Fig. 2.3). The fact that gravitational acceleration varies with latitude was not known when pendulum clocks first came into general use. The effect on pendulum clocks taken to different latitudes was the first firm indication of the latitudinal variation of gravitational acceleration. This variation indicates such a wealth of ideas about the earth, that its discovery was an important advance in the scientific study of the earth. Therefore, the pendulum occupies an important place in the history of the earth sciences.

The term seconds pendulum, appears frequently in the early literature on the earth's shape and gravity. This happened because the early pendulums which revealed the latitudinal variation of gravity were clock pendulums.

By a seconds pendulum, we mean one with its length so
adjusted that it oscillates with a period of one second
as was common in pendulum-controlled clocks. A pendulum
so adjusted for high latitudes will, as already indicated,
beat more rapidly if taken to the equator and will have
to be lengthened to become a seconds pendulum again and
vice versa.

At the time of the inception of the era of
ellipsoidal geodesy, the behaviour of the seconds pendulum
was a matter of widespread interest. As a background
to this question, Isaac Newton stated in a latter written
to Edmund Halley in 1686 that about 1671 "...I calculated
the force of ascent at the equator, arising from the
earth's diurnal motion, in order to know what would be
the dimunition of gravity thereby" (CAJORI, 1928, p. 172).
A few years previously, Christiaan Huygens had, in his
Horologium Oscillatorium, 1673, (SOCIÉTÉ HOLLANDAISE,
1932b, pp. 68-436) published the first correct theory
on centrifugal force, and, in the same publication, had
formulated the period of a pendulum in terms of gravitational acceleration as well as of its length. Thus the
theoretical basis was laid for expecting that the rate
of vibration of the pendulum would vary from equator to
pole. Huygens in fact knew of these relationships for some
years before publication of Horologium Oscillatorium:

he had communicated the expression for the period of a
pendulum to the Royal Society in anagram form in 1669
(SOCIÉTÉ HOLLANDAISE, 1932a, p. 246), the solution of
which he kept secret until 1673. Furthermore he had
calculated the effect of the earth's rotation on the period
of the pendulum as early as 1666 (SOCIÉTÉ HOLLANDAISE,
1932a, p. 285). However, he evidently still believed,
some years later, that the effect would be negligible, as
he advocates 1/3 of the length of a pendulum vibrating
seconds as a standard unit of length in Horologium
Oscillatorium, 1673 (see e.g. TURNBULL (ed.), 1960, v.
2, p. 312).

These exciting theoretical ideas, held by leading
scientists, inspired a widespread interest in the experimental
proof of such ideas and in the exploration of their further
consequences. From the standpoint of scientific theory,
therefore, the behaviour of the pendulum was, at the time,
a matter of the greatest importance.

The practical possibilities of the pendulum were
no less effective in generating interest in its application.
Christiaan Huygens designed the first working pendulum
clock, in 1657. This clock brought in a new era of accuracy
in timekeeping. He listed the following applications of
his invention:

> (1) To find the difference between the meridians
> with greater accuracy.

(2) To measure time more accurately than the sun.

(3) To be a perpetual and universal measure (quoted
       by GUYE and MICHEL, 1971, p. 102).

This last aim was one which was widely held before
it was realized that the period of a pendulum would vary
with position on the earth's surface; it was hoped that the
length of a seconds pendulum would remain sufficiently
constant over the earth's surface to serve as a universal
standard of length.   The Paris Academy of Science (of
which Huygens was an active member from 1666, the year
of its foundation) was, in the mid-seventeenth century,
interested in this possibility. This interest was one
factor which lead Jean Richer in 1672-73 to investigate
whether or not there would be a change in the rate of
vibration of a seconds pendulum which had been taken from
Paris to Cayenne.

However, the strongest incentive to investigate
the effect of location on a pendulum came from the timekeeping role of the pendulum clock. Expanding maritime
interests led to the need for accurate timekeeping in
distant parts of the earth in the work of extending
stellar catalogs, needed by navigators, and in attempts to
use chronometry for the determination of longitude. All
of these interests put questions raised by the pendulum into
the first rank of scientific problems. The close ties

of these problems with the earth's shape and gravity led
to the early development of the related branches of earth
science. The early ties of all these matters with clocks
and timekeeping lead us to at least a brief consideration
of this subject.

Clocks and timekeeping

Mechanical clocks in Europe were first devised in the
thirteenth century, although similar developments in
China preceded those in Europe by several centuries
(PRICE, 1959, p. 111). The first European clocks
were driven by a falling weight, which remained the principal
driving mechanism until spring-driven clocks appeared in
the fifteenth century. The rate of drive was from the
first controlled by some form of escapement which allows
the systematic escape of power through the driving
mechanism at a rate determined by an oscillator. The
oscillator was at first of the verge type. Clocks in
this period were commonly in error from one half to one
hour per day. In the late Middle Ages clock movements
became smaller and smaller, and at the beginning of
the sixteenth century, the watch appeared. The pendulum
was introduced to clockwork in 1657, and with improved
escapements soon proved to be capable of being accurate
to the nearest minute or better. The invention of the

oscillator balance spring for watches by Robert Hooke in
1658 (followed by further developments by Christiaan
Huygens in 1675) was an advance in watchmaking comparable
to the introduction of the pendulum for clocks, making
it possible to reduce errors in the timekeeping of watches
to the 5 - 10 minutes per day range (GUYE and MICHEL,
1971, pp. 11-18, 43, 104, 107). At this point we should
note the direct connection between practical concerns of
the age and the development in timekeeping. This came
in part from the problem of determination of longitude.

The longitude problem

In an age obsessed with overseas trade and access to trade
goods in distant lands, the fixing of points on maps and
charts and determination of the position of a ship at
sea were matters of the greatest concern. Latitude was
not a problem. It can be found by measuring the angular
height of the sun or stars above the horizon. Longitude
was more difficult, even though several astronomical
methods had been devised for determining it. These often
required better astronomical skill than was possessed by
the average practical navigator, and in general, the accuracy
obtained was not sufficient. In 1558, Gemma Frisius, a
Dutch astronomer and mathematician, proposed a method which
was extremely simple in principle. His suggestion was:

"people are beginning to use small clocks called watches and they are light enough to be carried...they offer a simple means of finding the longitude. Before setting out set your watch exactly at the time of the country you are leaving...When you have moved a certain distance calculate the hour of this place with the astrolabe; compare this with that of your watch and you have the longitude" (quoted by GUYE and MICHEL, 1971, p. 141). This method was to have continuing appeal because of its simplicity and because of the fact that its successful application was seen to depend only on technical improvements in the clock. Various governments recognized the advantages they would gain if they were the first to solve the longitude problem. In the seventeenth and eighteenth centuries, the governments of Spain, the Netherlands, England and France offered money prizes for the successful solution of the problem. The longitude problem occupied the attention of a long succession of scientists, including Christiaan Huygens, Isaac Newton, and Robert Hooke, with a considerable impact on science (MASON, 1953, pp. 99-212). In England, the Royal Society devoted considerable attention to the problem, as did the Paris Academy of Sciences in France. The Paris Academy of Sciences established its own observatory for the study of the longitude problem, and the observatory

was built during the period 1667-72. In England, the
Greenwich Observatory, which was to figure largely in
geophysics in later centuries, was built in 1675-6, for
the same purpose, and the post of Astronomer Royal was
established. The longitude problem was not in fact
solved satisfactorily until after the Scientific Revolution,
using Frisius' method, by LeRoy in 1763 and Harrison in
1765.

Many attempts using a great variety of methods
were made and some of these, even though failing in their
immediate objective, were of consequence for the future
science of geophysics. The study of the earth's shape
and gravity was advanced considerably because timekeeping
was recognized as the principal element in any successful
treatment of the longitude problem. For example, the
mechanics of pendulums and their operation at sea was of
much concern as was their rate of vibration and its
variation with position on the globe. We have already
noted the stimulus given to geomagnetism by attempts to
apply magnetic declination (Sec. 2.5) to the determination
of longitude at sea.

Gravitation and clocks

We now have considered a number of examples of how practical
applications of clocks and timekeeping bind these matters

to earth science. There is another connection, lying in
the very mechanism of clocks. The rate of drive in the
case of falling weights, and the time-constants of the
oscillators which control the rate of advance are, to
varying degrees, related to the local value of gravitational
acceleration.

The possibility of using this connection between
clock rates and gravity was recognized early. Francis
Bacon in his Novum organum (1620) suggested that the
difference in clock time between steeple and mine would
prove whether or not the attraction of the earth really
causes weight (CROWTHER, 1942, p. 123). However, sufficient
reliability and accuracy to detect differences in gravitational acceleration at different latitudes or different
elevations is not in fact feasible except in pendulum
clocks. The latter did not reach the required level of
development until a few years after their introduction.

Development of the pendulum

The isochronism of the pendulum (that is, the fact that
a pendulum of a given length has a fixed period of
oscillation regardless of its mass and, at least for small
oscillations, regardless of the amplitude of its oscillation) was first discovered by Galileo Galilei in 1583.
This principle is the basis of the application of the

pendulum to regulate the rate of running of clocks.

Galileo is said to have discovered isochronism in 1583 while observing a swinging chandelier in the Cathedral of Pisa (SEEGER, 1966, p. 170). It is a matter of interest that such an observation, of a type that conceivably could have been made at any time since ancient times, had in fact not been made, or a least had not been recorded, previously. The timing of the discovery may have been controlled primarily by the development of ideas. Galileo was only nineteen at the time of his discovery, and had only two years previously entered the University of Pisa, presumably to study medicine. He had, however, become interested in natural philosophy and likely had become acquainted with mathematics at the time of his discovery of isochronism (SEEGER, 1966, p. 4). It is likely that the discovery became possible because new ideas were abroad. Galileo's greatest scientific achievements later in life were possible because he had absorbed a new spirit in science and mathematics which had in his time begun to make itself felt. His achievements in mechanics, in which he laid the foundations of the laws of motion, were considered by BERNAL (1965, p. 296) to be possible because he was a master of "the new mathematics (symbolic algebra and trigonometry) that had blossomed with the Renaissance". His ability to find significance in a

simple phenomenon involving the quantitative description
of motion, which had apparently gone unnoticed before,
may more than anything have come from the influence on
him of the new ideas of the Scientific Revolution that
were springing up about him.

Even after the discovery of the isochronism of
the pendulum, the application of this device to clockwork
was not immediate. Galileo himself initially attempted
to apply the principle to a device for counting units of
time, and he approached Dutch seafaring interests to
adopt a pendulum-counter for determining longitude at
sea. Also, astronomers employed Galileo's development in
the form of pendulum time-counters (LLOYD, 1957, p. 662).
The idea of applying the pendulum to clocks occurred to
Galileo about 1641, although he did not himself complete
a working clock. He began one which was partially completed
after his death by his son in 1649. The first working
pendulum clock was built in 1657 by Salomon Coster,
following designs and instructions supplied by Christiaan
Huygens, who continued afterwards to develop the pendulum
and pendulum clocks with one of his objectives the
determination of longitude at sea. Errors in timekeeping could, with these new clocks, be held down to a
few minutes per day. The development of the anchor
escapement in 1670 led to sufficient reliability and

accuracy (errors less than one half minute per day
(GUYE and MICHEL, 1971, p. 27) to make it possible for
pendulum clocks to detect the variations in gravity that
occur from place to place on the earth's surface through
their effects on timekeeping.

Gravitational effects on pendulums

From our vantage-point at the present time, a summary
of the ways in which the period of a pendulum may be
influenced by variations in position and altitude is
given in Figures 2.3 and 3.10.

Two expeditions to equatorial regions

Richer in Cayenne (1672-3) - One of the earliest proposals
(made early in 1667) to be considered by the newly formed
Académie Royale des Sciences was one put forward by
Adrien Auzout, an astronomer and a leading spirit in the
establishment of the Academy, that an expedition be sent
to equatorial regions to carry out astronomical observations. The Academy began to discuss the idea, as
stated by its Secretary, of "sending observers under
the patronage of our most munificent King into different
parts of the world to observe the longitudes of localities
for the perfection of geography and navigation". (OLMSTED,
1942, p. 120). The proposal was supported by the government,

and the expedition set sail in 1672 for Cayenne in French Guiana. The aims of the expedition were to use certain advantages of an equatorial location to improve existing tables of solar and planetary motions, and to observe the positions and magnitudes of southern stars not visible from Paris.

These aims were considered important by the Academy and the government for several reasons. First, a revolution in observational astronomy between 1650 and 1670 had made new observations desirable as part of a new era in astronomy springing up following the technical advances. We have already noted that a general advance in instrument making set in about 1650, as a result of the Scientific Revolution (Sec. 2.5). Second, the general spirit of the age, in which maritime expansion was felt by a country like France, for which its extensive trade and overseas connections were of primary importance, favoured any extension of knowledge that could advance navigation and other geographical concerns in distant regions.

Because the question of whether or not the rate of oscillation of a pendulum varies with geographical location was recognized as a fundamental one by many of the members of the Academy, the investigation of this matter was included as one of the aims of the expedition. It should be emphasized, however, that the principal aim was astronomical observation.

The importance of the pendulum results was not recognized, at least by many of the academicians. In fact, the majority of members of the Academy did not at first credit Richer's discovery that his pendulum clock, regulated at Paris, lost time in the equatorial zone. They apparently expected and hoped for a world-wide constancy in the pendulum rate, which through the length of a seconds pendulum, could become a standard of length. The expedition made an important contribution to astronomy and was the first extension of European astronomy to equatorial latitudes following the evolution in instrument design (OLMSTED, 1942, pp. 123-124). For earth science, however, the observation regarding the length of the seconds pendulum was the expedition's most important discovery. It was found that Richer's pendulum, adjusted as a seconds pendulum in Paris, lost two and a half minutes per day in Cayenne. It had to be shortened to restore it to correct timekeeping, and the amount of shortening required was carefully recorded. The fact that it was recorded is of interest when compared with Halley's experience with his pendulum clock, to be discussed in the next section. The recording of data on the pendulum adjustments was made because the level of interest in Paris regarding this phenomenon. There was not sufficient theory to provide a clear guide on what might happen, although interest in the results was sufficient to alert

the expedition to the possible importance of the measurement and therefore to the necessity of taking care about
these observations in case they should prove important.

It is interesting to note that notwithstanding the
above, the significance of Richer's results was not generally
understood for some time. Dr. Papin (a member of the Paris
Academy) remarked in 1680 to the Royal Society of London
"that a person employed by the Royal Academy of Sciences
at Paris to try pendulum clocks in places near the line
found them to go much too slow; and that the said Academy
doubted the truth of this fact, but supposed, that he had
been some way mistaken, though he with much confidence
affirmed the matter of fact to be true, but knew no reason
of it" (BIRCH, 1756-7, v. 4, p. 1).

A second expedition, sent independently to equatorial
regions to make astronomical observations, encountered an
effect similar to that found by Richer. The observer, Edmund
Halley, was apparently unaware at the time of Richer's similar
experience. He therefore made an independent confirmation
of the earlier result obtained at Cayenne. Apparently
Halley did not publish his result for several years later,
although he had a conversation with Robert Hooke about
it soon after his return to England.

Halley at St. Helena (1677-8) - In 1677, Edmund Halley
began work on a Catalogue of the southern stars, based on

observations at the island of St. Helena. Richer had
made important observations at Cayenne, which formed a
creditable but not extensive catalogue of southern stars.
Halley's purpose in going to St. Helena was to make a much
more comprehensive catalogue using, as had Richer, newly
developed instruments. Like Richer, he employed a pendulum
clock for his astronomical observations. His clock was
adjusted to beat seconds at London, and (similar to Richer's
experience) began to lose time after transport to the
equatorial regions. Apparently Halley had not been aware
of the possibility of a planetary variation in the period
of a pendulum. He remarked later in a paper on the nature
of gravity (HALLEY, 1686, p. 7) that "the length of the
pendulum vibrating seconds of time is found in all parts
of the world to be very near the same. T'is true at
St. Helena...I found that the Pendulum of my clock which
vibrated seconds, needed to be made shorter than it had
been in England by a very sensible space, (but which
at that time I neglected to observe accurately) before it
would keep time; and since the like observations had been
made by the French Observers...yet I dare not affirm
that in mine it proceeded from any other cause, than the
great height of my place of observation..."

    This lack of awareness of a possible latitude
effect on gravitational force (which, as we know today,

results in the time effects shown in Fig. 2.3) is not surprising, since the expeditions of Richer and Halley were the first in which this effect could have been clearly observed. Theory had not developed to the stage (although it was soon to do so) of giving a clear indication of what was to be expected.

Vertical gradient of gravity

It is interesting to note that another line of development in ideas had been going on for some time, and Halley's first reaction to his results is of interest in this regard and may have been influenced by it.

The existence of a vertical gradient of gravity had interested scientists in England since the time of Francis Bacon. Attempts in the seventeenth century to measure this gradient reached their peak in a series of experiments by Robert Hooke. The principal experiments were in 1662 and 1666 when he attempted to detect differences in weight of objects between Paul's Steeple and the top of Westminster Abbey and the ground, and between the surface and points in deep wells (BIRCH, 1756-7, v. 2, pp. 69-73; v. 1, p. 164) using a balance and long wires or packthreads to suspend weights at the desired elevations. These measurements were not conclusive, and Hooke recognized that "we must have some way of trial much more accurate, than this of scales" and suggested

that one possible method is "by the motion of a swing-clock:
for if the attraction of the earth towards its centre be less,
the farther the body is placed above or below its surface,
then the motion must be slower there than when placed on the
surface" (BIRCH, 1756-7, v. 2, p. 72).

Hooke's interest in this matter is shown in a letter
written to Isaac Newton in 1679 (TURNBULL, ed., 1960, v. 2,
p. 310), where he describes his reaction to Halley's account
of his St. Helena experience, made to him in a conversation
shortly after the latter's return. Hooke immediately saw in
this result the effect of the vertical gradient of gravity,
explaining the retardation of Halley's pendulum clock as being
due to the decreased gravitational force that would be found
at Halley's observation point as a consequence of its height.
It was on a mountain at considerable elevation (2400 feet) above
sea level; it would, as Hooke saw, be farther from the earth's
centre of attraction than if it lay at the base of the hill,
and therefore, would have lower gravitation. No mention was
made at that time of the possible effect of latitude (even
though this was, as we now know, the principal cause of the
effect on Halley's clock). Hooke had speculated in 1666 on
the effect of centrifugal force (one of the two main causes
of planetary variation in gravitational force) (BIRCH, 1756-7,
v. 2, pp. 90-92), and had given an illuminating picture of
planetary motion based on his characteristic experimental

approach. In putting forward these ideas, Hooke made a major contribution to the theory of gravitation. They contributed to Newton's _Principia_ and to the general acceptance of the Newtonian view of gravitation. There was not at that time any suggestion of a latitude-controlled variation of gravitational force.

This suggestion did, however, come from Hooke in 1680, only one year after his conversation with Halley regarding the latter's experience at St. Helena. He remarked then that a latitude effect on the pendulum might occur. At a meeting of the Royal Society, as recorded by BIRCH (1756-7, v. 4, p. 1), "Mr. Hooke read...(an)...account of his theory of circular motion and attraction...and deductions from that theory; as (1) That pendulum clocks must vary their velocity in several climates. (2) That this variation must also happen at different heights in the same climate: Which last remark he confirmed by an observation of Mr. Halley at St. Helena; and (3) As a consequence of these, that a pendulum was unfit for a universal standard of measure."

After Isaac Newton's development of the theory of gravitation, it was recognized that the dominant effect in Richer's and Halley's experience was the latitude effect. The vertical gradient in gravity, however, was still an important effect, and its significance was felt at a later date. The gradient proved to be related to a planetary quantity,

the earth's mean density (hence indirectly to the mass of
the earth). The discovery of this relationship by Pierre
Bouguer and La Condamine (Sec. 3.7) initiated a new earth
science and began a long series of determinations of this
gradient.

The theory of gravitation, a development towards earth science

A quantitative theory of gravitation finally came into
being through the work of many scientists in the latter
part of the seventeenth century. This work is perhaps
best exemplified in the published works of Christiaan
Huygens and Isaac Newton. Both were keenly interested in
Richer's results and used them as illustrations of their
theories regarding terrestrial gravitation. Huygens
had, as a founding member of the French Academy, which
in its early years had sent Richer to Cayenne, a particular interest in the results of that expedition. Supposing
a spherical earth and centrifugal force due to the earth's
rotation, he calculated that a Paris seconds pendulum would
have to be shortened by 1/580 of its length to beat at the
same rate in Cayenne. This was a remarkable calculation for
its day, as it explained correctly the sign and the general
magnitude of the effect.

The required shortening applied by Richer was 1/380

of his pendulum's length. The discrepancy between 1/580 and 1/380 resulted because the flattening of the earth had not been taken into account.

Newton, by including the flattening, successfully explained Richer's results. In his _Principia_ (Book III, Prop. XX, Problem IV) he gave the first completely satisfactory explanation of Richer's experiment. He assumed the earth to be an oblate spheroid in hydrostatic equilibrium under gravitational and centrifugal forces, and calculated the effect on the seconds pendulum of both these forces. The agreement between his calculation and Richer's result may be taken as an early example of the fact that the shape assumed (oblate spheroid) and Newton's gravitational theory were correct. Thus studies of the earth played a considerable role in the establishment of Newton's theory of gravitation.

Further, he derived a first order approximation to the law of variation of gravitational acceleration with latitude, finding that the acceleration varies as the square of the sine of the latitude. He also calculated the flattening, assuming that the earth is homogeneous, to be 1/230. This was the first of a long line of calculations of this most important quantity, the value which has a very significant bearing on the internal constitution and history of the earth. The value has

been gradually refined during subsequent centuries, and
the most recently determined value (from satellite geodesy)
is 1/298.247. Often far-reaching conclusions about planetary interiors can be drawn from the values of a single
planetary parameter such as the flattening, when coupled
with the appropriate scientific theory. For example,
Newton deduced from his proof of flattening that the
earth had once been a rotating fluid body.

The density of the earth's interior is another
physical property value which can be deduced from a few
very general determinations. Isaac Newton in the Principia
established from considerations regarding the solar
system that the mean specific gravity of the earth
is between 5 and 6. He concluded that given the known
densities of surface strata (overwhelmingly less than 5),
the earth must be much denser at depth. Such deductions,
of which many important examples can be found in the
study of the earth, are of considerable interest. They
are based on the results of observations which, although
often skilled and imaginative, are fairly simple when
considered against our scale of sophistication in observation and experiment (Sec. 1.5). However, when combined
with an ingeneous application of basic scientific theory,
they produce remarkable advances in understanding. This
approach is intertwined as one of the threads with the

others leading to modern earth science. In contrast, a second approach to the problem of internal density, more linked with experiment, was developing.

Francis Bacon had proposed an experiment to determine whether or not there is a vertical gradient in gravitational attraction. In the early days of the Royal Society, experiments along these lines were carried out by Robert Hooke. These approaches represented a way of studying gravitation, different from those used in the Newtonian theory. Such approaches are closer to true experiment, and thus higher on our scale of sophistication of observation and experiment. It is another thread of development which we shall follow through the development of earth science. The next step after Hooke's experiments were those carried out by Pierre Bouguer in 1737 and by himself and La Condamine the following year to measure the attraction of a mountain. Bouguer published the rather interesting theoretical result that such a measurement does not merely give a measure of the gravitational attraction of the mountain, but leads also to the determination of the mean density of the earth. The place of this discovery in the development of the earth science is discussed more thoroughly in a later section (Sec. 3.7).

Subsequent developments in the theory of the Figure

of the Earth came more slowly. Isaac Newton had assumed
that the earth is an oblate spheroid in explaining
Jean Richer's pendulum observations. His development
showed another significant possibility regarding the
earth, that the earth's shape must also be the equilibrium form for a rotating, gravitating fluid. If correct
this fact would suggest that the earth was once fluid,
or if it was always solid, that it is capable of adjust-
ment by means of creep or plastic flow to long-term forces.

Further developments came a few decades later,
when James Stirling and A. C. Clairaut, developed
improved expressions for the equilibrium form of a
rotating, gravitating fluid in 1737. Colin Maclaurin
published a satisfactory proof that an oblatum is an
equilibrium form in his prize essay on the tides in
1740. The further problem, to show that an oblatum
is the only possible form for a slowly rotating body
such as the earth, was not satisfactorily solved until
1784, by Legendre.

The reason for such a long delay in the solution
of this problem (in spite of the fact that at the time
there was a great deal of interest in this and related
problems) was largely technical. Isaac Newton and his
immediate successors had developed the integral calculus,
which was adequate for the simpler forms of the problem.

For Legendre's solution new mathematical techniques were required. These were developed, in the form of spherical harmonics, by Legendre and also independently by Pierre Laplace. It is interesting to note that these techniques, applicable to a great range of scientific and mathematical disciplines, were originally developed in the course of the solution of earth-directed problems.

2.7 The geological sciences during the Scientific Revolution

The geological sciences did not come into the forefront of science until a later period, the Industrial Revolution, when they experienced a major phase of growth. However, the forerunners of these sciences went through a definite (although smaller) growth phase during the Scientific Revolution. This was primarily due to an accelerated search for minerals, brought about by an increased demand for raw materials (including metals) for the manufacturing industry, which began to grow with the increased economic activity of the period. As BERNAL (1965, p. 220) remarks, during the Renaissance the greatest advances in technology were in the "closely linked fields of mining, metallurgy and chemistry. The need for metal led to the rapid opening of mines first in central Germany and then in America".

      The mining tradition of the Freiberg district of

Saxony (dating from at least the fourth century) inspired
Georg Bauer (latinized to Agricola) to publish what were
to become two of the most important early books on earth
science, De natura fossilium (1546) and De re metallica (1556).
Agricola has been referred to as "the father of modern
mineralogy" (ADAMS, 1938, p. 185). This scientific tradition,
derived from mining, continued at Freiberg. It led later to
important contributions to the development of the geological
sciences, which occurred during the Industrial Revolution
(Sec. 4.4). In De Natura Fossilium (1546) Agricola put forth
a system of mineralogy. Although imperfect on present-day
standards it represented a sufficiently significant advance
in the subject for a present-day writer to say: "it was
here that the science of mineralogy, in the modern acceptance
of the term, took its rise" (ADAMS, 1938, p. 175).

This interest in minerals gave rise to a number of
developments including the concept of the geological map,
the scientific description of fossils, interest in the
processes of sedimentation and in structural geology, and
the rudiments of stratigraphy (SCHNEER, 1954, pp.266-7;
1967, p. 5). In addition, interest in the earth generated
from the great navigations gave rise to many theories of
the earth, which touched on geological subjects. Robert
Hooke's Lectures on Earthquakes (1688, published posthumously
in 1705) was a remarkable work on historical geology, almost

a century ahead of its time.

However, a major growth of the geological sciences
did not begin until the mid-eighteenth century, at the
opening of the Industrial Revolution (Sec. 4.4). The
major threads and their interrelationships in the development
of the geological sciences are shown in Fig. 4.1).

2.8 The close of the Scientific Revolution

The early eighteenth century pause

Science declined seriously in most of Europe in the early
eighteenth century. The basic stimulus for the higher
level of activity preceding this period (the commercial
expansion of the sixteenth and seventeenth centuries)
had dropped off and science did not show renewed vigour
until the agrarian and industrial revolutions of the second
half of the eighteenth century (MASON, 1953, p. 224;
BERNAL, 1965, p. 358). The beginning of the seventeenth
century marked the end of the great upsurge of science
during the Scientific Revolution. Modern science and the
accompanying scientific ideas had become established,
ushered in by that great age of science. The drop in level
of scientific activity was followed by a pause in science
which lasted until the second half of the century.

There were, of course, differences in the time and

rate of the decline from one country to another, depending on their particular histories and conditions. In France the scientific thrust continued into the seventeenth century, for reasons to be examined in a later section (Sec. 3.5).

2.9 Summary and plan

We have traced two threads of science through the Scientific Revolution: terrestrial magnetism and gravity and figure of the earth. The first was seen to grow into an earth science before the end of the period. The second was seen continuing on into the following century, with indications that it, too, would consolidate into an earth science. These examples were featured in the present chapter because they are particularly good illustrations of the fact that scientific history is closely linked with social and general history. Having seen this link illustrated in the present chapter, we will go on in the next to consider ways of studying growth quantitatively. These methods will then be applied to the scientific study of the earth in order to understand the pattern and meaning of its growth.

Chapter 3

QUANTITATIVE STUDY OF THE GROWTH OF THE EARTH SCIENCES

In the opening chapter it was stated that, in common with
all branches of science, the earth sciences have developed
in a sequence of stages. For a given branch of earth science,
the definitive stage is found to be one of rapid growth
centred on a relatively short period of time. It was
pointed out that these periods coincide with decisive
periods in the history of science as a whole, and in general
history. The Scientific Revolution is one such period, as
is the Industrial Revolution. We have traced a number of
branches of earth science (or their forerunners) through
the Scientific Revolution. We have seen how different component threads of these sciences, and the sciences themselves,
have followed their various paths as part of the development
of science in particular countries and of the shifts in
scientific initiative from one country to another. They
have interacted with other sciences and with the social and
technological demands of the age. Having seen these developments in some detail, we are now in a position to examine the
growth of earth sciences using a quantitative analysis.
As stated in Chapter 1, the history of geographical discovery
is an appropriate beginning for such an analysis, since in

the Scientific Revolution the earth sciences arose largely
in response to problems posed in the extension of trade
and navigation to the global scale.

3.1 Geographical discovery since the beginning of the Scientific Revolution

The use of quantitative indices

The question whether the analysis of simple indices such
as used here can ever adequately reflect the complexities of
social development is often raised regarding quantification
of social and historical data. This argument can be developed
for geographical discovery. There are many factors which
can be pointed out as controlling the direction of development of geographical enterprises. Economic, military, and
imperial aims all control the vigour and direction of geographical enterprise. These factors intertwine in many subtle
ways, and it might be argued that at most only a small fraction
of the significant influences can be captured in numbers or
quantitative indices. Because of these circumstances, poorly
chosen indices might portray only certain aspects of the development. If the limitations of such indices are not clearly
emphasized, misinterpretations of the indices could be made.

One strongly expressed affirmation of doubts of this
kind was made by DEDIJER (1968, p. 19), in a discussion

which happened to be related to the application of growth curves to the development of medieval universities, in his statement that "all attempts to draw logistic or other curves through scattered data represents an unfortunate attempt to make social phenomena behave like those in some parts of nature and obfuscates the difficulties in making an objective social science study of the growth of medieval universities or any other social phenomenon."

There are, on the other hand, examples of the successful application of the quantitative method. As one such example we may cite the results of LILLEY (1948, pp. 180-231) who in the first edition of his book on the history of invention uses a simple index to produce graphs of the rate of invention, and of its accumulation. From these curves he came to significant conclusions regarding the interactions among general history, economic development, and invention. This example is particularly interesting because it also shows some of the difficulties in quantitative work. In a second edition of his book (LILLEY, 1966), the author added sections to cover inventions made in the interval since the first edition. In the second edition he declined to include a graph of the rate of invention, because the flood of invention since the first edition made the quantitative analysis of the new material an impossible task for himself as an individual author revising an earlier work, which would be "prohibitively heavy". He expressed his belief in the soundness of his earlier approach, saying that "it remains my hope that my early essay in this direction will not be forgotten" (1965, p. xii).

Anyone feeling uneasy about growth curves should first review the descriptions of their applications, which include systems ranging from biological organisms to components of sophisticated systems created by human society, such as are given by DAVIS (1941, pp. 209-241) and by PRICE (1961, 1965). In the last-mentioned book it is amply demonstrated that simple indices can lead to profound conclusions about the development of science. The author showed that a logistic growth curve very often describes the course of scientific development. In such a growth process, a slow initial accumulation grades into exponential growth. Finally factors enter which check the growth and cause a levelling off of the total accumulation. The exponential growth phase resembles many features of an epidemic (CROWTHER, 1967, p. 302, 304). Some particular phases of development in science can be portrayed reasonably well by the epidemic model, and a mathematical analysis of growth based on the theory of epidemics can throw much light on the development of science in these particular cases. GOFFMAN (1971) successfully explains the growth of symbolic logic from 1847 to 1962 on the basis of this model.

CROWTHER (1967, pp. 299-306) while recognizing the importance of growth curves as applied by de Solla Price, the use of which he describes as "one of the most striking contributions to the history of science since the Second World War", also points out the dangers of such an approach. He notes that interest in quantitative methods also serves to divert attention "from qualitative to quantitative questions". In history of science, where an understanding of the interrelations among scientific development and economic and general history is of the utmost importance, complete diversion from these chiefly qualitative considerations would be disasterous. As he said about a possible negative effect of the introduction of computers, (CROWTHER, 1967, p. 170) "When calculations can be facilitated, there is a temptation...to substitute industriousness in calculation for the study of significant questions and matters of qualitative and primary importance". As an example of important qualitative approaches, the same author (CROWTHER, 1967) has shown the importance of biography as an element in the effective treatment of the history of science.

On summing up all of these points-of view, we may, I think, conclude that quantitative indices can be applied as an important but not the sole element of investigations in the history of science. Their full potential, however, can not be realized unless they are integrated with the

equally important qualitative elements without which the
history of science would become a sterile exercise.

Let us approach the problem of the appropriateness
of a quantitative analysis of the development of geographical
discovery by making a trial of the method. We will begin
by looking at the problem and laying down the terms of
quantification, and selecting the indices to be used.
We will then examine how these indices change with time,
to see if the picture thus revealed is a reasonable one.

A quantitative analysis of the growth of geographical discovery, 1450-1900

The graph on Figure 3.1 shows the frequency of land
expeditions or sea voyages sent out or directly sponsored
by various countries. This emphasis, on the origin of an
expedition rather than the territory surveyed by it, is in
accordance with the fact that throughout the period
studied, scientific, technical and related developments
have tended towards centralization in a small number of
centres (BERNAL, 1965, p. 486, 968, 971). Geographical
discovery has been to a large extent motivated by the
dynamics of these centres. We shall see a correlation
between the growth of the earth sciences and that of
geographical discovery when treated as part of the dynamics
of centres of growth. For our immediate purpose (examining

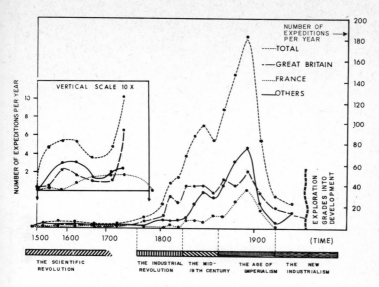

Fig. 3.1   Rate of accumulation of geographical
expeditions.

the growth of earth sciences), the fact that a given nation
sent out an expedition is more important than is the particular territory on which it operated. The other side of the
question: how the areas of operation spread across the globe,
or how geographical discovery developed in a given area
cannot be ignored and is an interesting and important question
in itself. Of course it does have some influence on developments in the initiating centres themselves, and the growth
of earth sciences has been influenced by the way in which

the course of exploration proceeded in particular areas.
The history of the long series of voyages to the Canadian
Arctic and of the Russian advance into Siberia, through the
scientific programmes that accompanied them, influenced the
development of many branches of earth science. For the rather
general analysis of the present book, however, these questions
are of lesser importance than they would be in more detailed
localized studies.

A single comprehensive history of geographical
discovery, one which is recognized as an authoritative
study, that by BAKER (1937), was chosen as the source of
material. It was hoped that by choosing a single
comprehensive history, a maximum degree of uniformity in
comprehensiveness and detail would be maintained. Only
expeditions which were chiefly for the purpose of geographical discovery or mapping were used. Included were
some primarily commerical enterprises (such as whaling
cruises) if mapping was in fact one of its aims. Expeditions
for purely commercial, political or adventurous purposes,
if they did not result in any substantial contribution to
geographical knowledge relating to the physical configu-
ration of the globe, were omitted. Expeditions satisfying
these broad criteria were given a score of one each.
It was decided not to apply any value-judgements beyond
the minimum criterion of producing a mentionable contri-

bution to geographical science. Thus, for example, no
expedition was given extra weight for what appeared to
be an outstanding contribution to geography. Nor were
varying lengths of time involved given different weights,
beyond the fact that if an overall effort lasted for more
than ten years and if this period consisted of several
producing phases, each phase was given a separate count.
The total count for fifty year intervals up to 1800, and
for ten year intervals after was made and the number of
expeditions per year for each interval was calculated.
Sub-totals, and corresponding values of the total were
also obtained for several countries. Values were then
plotted on a logarithmic time scale, at the centre of each
interval.

The rate of exploration and its variation with time

The plot of the number of expeditions per year against time
is shown in Figure 3.1. Examining the graph for the world
total, we see first of all a rather interesting pattern.
The curve consists of a number of overlapping peaks. As a
comparison between these peaks and the time-scale shows,
these peaks of geographical activity are centred upon major
periods in the history of science.

We have seen that from a general analysis of the
history of science, six important phases are discerned

since the beginning of modern science. These are: the Scientific Revolution; the Industrial Revolution; the Mid-Nineteenth Century period; the Age of Imperialism; the New Industrialism, Phase 1; and the New Industrialism, Phase 2. Furthermore, general histories of science show that these phases are also important periods in general history. Figure 3.1 does not extend into the age of the New Industrialism, and thus the correlation noted above cannot be tested for this period. The time span represented in the Figure ends just before the beginning of this period, because this point in time marks the beginning of a new phase in history following the general spread of human occupation across the globe. By this time, normal development becomes difficult to distinguish from exploration. A quantitative study in this later period is undoubtedly possible, but would involve more detail than is possible or desirable in a general overview such as the present book. Thus the Figure carries the notation "exploration grades into development", marking the end of the era of discovery and the beginning of one of development over virtually the whole of the globe. Thus our survey ends about 1900.

The shape of the graph of the number of expeditions per year versus time indicates a sequence of logistic curves in escalation, with the points of maximum growth centred on critical periods in the growth of science. This result suggests that a spirit of enquiry on a broad front, embracing geographical discovery as well as science, advanced in a series of waves with maximum intensity centred on a number of critical periods. It should be emphasized, as noted before, that these periods are divisions of general history as well as of the history of science. The fact that such a result appears in Figure 3.1 may be taken as an indication of the validity of our method of analysis. Given such an indication of validity it is appropriate at this point to consider ways of interpreting growth curves.

Principles of analysis of growth curves

In the preceding section, we have seen a good indication that growth curves can provide a reasonable picture of the development of geographical exploration. It is then appropriate for us to examine at this point the meaning of such curves and methods whereby the curves can be interpreted clearly and simply.

Any process of growth described by a logistic has certain features governed by the mathematical nature of the logistic growth curve. A brief summary was given in Sec. 1.1.. A more detailed treatment is given by DAVIS (1941) and by PRICE (1961, 1965). This type of growth is such that it is at first governed principally by a factor which is proportional to the amount of growth already achieved. This stage of the growth cycle is of the much-studied exponential type. However, with a logistic, a second factor gradually grows in importance with time. This factor inhibits the growth, and for a logistic curve in the exact sense, is proportional to the square of the amount of growth. Thus this latter process, growing faster than the earlier more nearly exponential process, finally damps the growth which therefore levels off and finally ceases altogether. Thus if a logistic can be shown to be the growth curve in any particular case, the nature of the processes acting to control it can be specified to some extent.

We must recognize of course that in science many similar models often fit the observed data equally well. Thus, we might, in our present analysis, be best advised to consider a logistic fit to the data as representing only the general features rather than the exact mathematical requirements of this type of growth. Thus we might more properly speak of a <u>logistic-type growth</u> as being applicable, and reserve the possibility of finer distinctions for future enquiries.

Individual scientific fields, in contrast to science as a whole, commonly develop with a logistic-type growth. The growth may, in some cases, follow a succession of stages centred successively on several time periods. LILLEY's (1948) growth curves for inventions are of this type. If a whole range of scientific fields are considered together, the episodic character of growth may be obscured, because different fields may peak at different times with the result that the individual peaks are not distinguishable. The observation of PRICE (1961, 1965), after considering the volume of published scientific work in all countries and its variation through time, is an example of the result of averaging in which he sees a steady exponential rise of science over the past three centuries. This is undoubtedly an important general conclusion regarding the growth of science. We will, in contrast, be considering single fields of science and their development in various separate scientific centres. Thus we are doubly restricting our view: to single fields, and to separate centres. The latter is a distinction of some importance, because science has always been, and continues to be concentrated about a number of scattered centres. We will, then, expect to see features of individual growth not outlined in more generalized studies such as those of de Solla Price.

We have defined in Sec. 1.1 the relationship between an index and its cumulative total (giving the amount of growth), as well as the critical points on these curves (Fig. 1.1). The "total" curve in Figure 3.1 represents three, and perhaps four, logistic-like growth curves in

escalation. The main features of the three principal
phases of growth are summarized in Table 3.1.

The critical points for the three phases of growth,
giving the times of the maximum rates of increase, all
correspond to the mid-points of major periods in general

Table 3.1

Numbers of geographic expeditions for scientific
purposes

	Period	Scientific Revolution	Industrial Revolution Mid Nineteenth Century	The Age of Imperialism
(1)	Dates	1450-1700	1750-1860	1890-1918
(2)	Accumulation during the period in the number of expeditions	120	950	150
(3)	Date at which maximum rise is centred	1560	1850	1890
(4)	Accumulation per year	1/2	8 1/2	5

history and in the history of science. Thus it appears
that each of these periods occasioned an upsurge of
expeditions, which died off as the periods came to a
close. The accumulation was greatest during the mid-
nineteenth century period.

As a confirmation of the validity of the quantitative method employed, we may compare the remark of BAKER
(1937, p. 489) in summing up his History: "The history of
geographical discovery and exploration shows no continuous
progress, but a series of advances followed by periods of
inactivity or of actual regression". This pattern of
growth, suggested by Baker, is in accordance with the
succession of growth curves resulting from our quantitative
analysis. Both, it is true, are based on the same body
of data (Baker's book). However, the author's criteria
come from a more qualitative view of the data, chiefly
embodying the impact and importance that the various
periods had on enriching the body of geographical knowledge.
We chose to avoid stressing this aspect of the subject.
In turn, he did not cite any counts of numbers of expeditions
or any similar index. Thus agreement between our method
and the author's summary, because of the different
approaches used, is therefore a confirmation
of the methods we are employing. BAKER (1937, p. 489)
gives further detail when he remarks that "there were

few startling discoveries in the seventeenth and in the first half of the eighteenth century". This circumstance is confirmed by our Figure 3.1, which shows a declining growth rate in exploration for this period. Baker notes the "outburst of exploring activity which reveals the secrets of the Pacific Ocean". This is marked on our Figure by the onset of the most rapid of the three phases of growth indicated by our analysis. Baker further notes the two distinct growth curves for the nineteenth century (Figure 3.1 and Table 3.1), remarking that "the most distinct turning point is found near the middle of the nineteenth century". The parts of the world being explored were quite different in the latter half of the century than they were in the first, and the type of exploration changed, with scientific explorers in the later period giving precision to the more general earlier work.

We may further note connections between the history of geographical discovery and economic history. A quantitative study of the rate of growth of overseas trade in Great Britain from 1688 has been made by DEANE and COLE (1962). There are some interesting connections between these authors' results and the graph for Great Britain on Figure 3.1. On p. 29 of their book these authors list the rate of growth of domestic exports plus retained imports from 1700 to 1910. There is initially a

slow rate of growth, accelerating markedly in the final
two decades of the eighteenth century. This acceleration
continues, with the peak of the growth rate of about 4
percent per annum in the two decades centred on 1850. This
is (if we smooth over a thirty-year span as was done with
the trade figures) exactly what our index of exploration
activity does. DEANE and COLE (1962) remark that the trade
indices follow closely many other indices of economic
activity. Thus we may conclude that changes in the
volume of geographical exploration were determined, at
least in the British case, by the course of development
of economic activity and overseas trade. In spite of
the steady drop-off of the rate of growth in overseas
trade after the mid-century, the authors note (p. 235)
that the shipbuilding industry continued to have a high
relative importance in the British economy until the 1880's.
Thus the general atmosphere (a country with a high level
of overseas trade and a continuing and vigorous shipbuilding industry) caused exploration to continue for
some time after the decline in trade began to set in.
They remark (p. 29) "it was in the 1880's in the period
of "the Great Depression" that the rising tide of British
exports was seriously challenged by foreign competition".
This is reflected in the sharp rise of the graph in
Figure 3.1 for "others" above the British curve. Geo-

graphical discovery on a large scale was brought to a close for all countries after 1900, because by this time unexplored territory was all but exhausted. At this point we can say: "exploration grades into development".

The shifting thread of initiative

Another feature of the growth of geographical exploration, besides the manner of its total growth, is the way in which the initiative in exploration shifted from country to country. This is also shown in Figure 3.1, where the number of expeditions per year is shown through time in its distribution among countries. Here we see a pattern of change extending through time and space. This pattern will be compared in a later section (Figure 3.3 and 3.4) with those for geomagnetism, mathematical studies of the earth's figure and gravity, and the geological sciences. This comparison, along with a discussion of science in various countries, illustrates another feature of the growth of the earth sciences: the shift of initiative from one country to another. This shift parallels that of history and of science in general. Trade and economic supremacy passed from one country to another, as shown for example in the Cambridge economic history volumes (POSTAN and HABBAKUK, 1966). The initiative was held first by Spain and Portugal. Then the shift was to Holland and later to

France and Britain. A part of this pattern is seen in
Figures 3.1, 3.6, 3.7 and 3.8. Finally, late in the
nineteenth century the index of geographical discovery
reached its peak with maxima centred again on Great
Britain and also on other European countries which had not
figured prominently in the earlier periods. This last
phase of activity was part of the scramble for colonies
which took place during the late nineteenth century.

This consistency among our quantitative indices
and their relationship to economic and general history may
be taken as further confirmation that simple indices can
be effective. Further detailed weighting of expeditions
and other refinements should be brought into future studies,
but are not required in a general overview as attempted in
the present book.

Relationship to earth science

The growth of geographical exploration fits a sequence of
logistic-type curves reasonably well. The maxima of growth
represented by these curves are centred on critical
periods in the growth of science. These critical periods
are also ones of importance in general history. In order
to explore the relationships among geographical exploration,
earth science, and general history, let us now (using
similar quantitative methods) examine the development of

some branches of earth science. We have seen the relationships among history of science, general history, and the
development of exploration. If we can establish a
connection between the history of earth science and any of
these, we will have placed earth science in its proper
historical context. The beginnings of scientific fields
which later formed geophysics have, in earlier writings,
been shown to lie in geographical exploration and related
sciences (TAYLOR, 1937; 1948). This is one connection
that we will explore in some detail.

3.2 A quantitative study of the growth of geomagnetism through the Scientific Revolution

We have followed (Sec. 2.5) the story of the development
of the early phases of geomagnetism. A quantitative study
of the growth of geomagnetism during the Scientific and
Industrial Revolutions was conducted as follows. Items
regarding measurements, expeditions, the development of
scientific instruments or techniques, new theoretical ideas
and the discovery of new phenomena were compiled from a
number of sources. An equal weight was assigned to each,
and the only criterion for importance of the item was the
fact that it has been included in the sources used. These
are authoritative references to the period covered, and

include the following: MITCHELL (1932, pp. 105-146; 1937, pp. 241-279; 1939, pp. 77-79; 1946, pp. 323-351), HARRADON (1943, pp. 3-17, 79-91, 127-130, 197-202; 1944, pp. 185-198; 1945, pp. 63-68), VON HUMBOLDT (1858), DE MOIDREY (1904), and NEEDHAM (1965a). Also included are other items referred to in the present book.

The growth is shown in Figure 3.2. Two principal phases of growth are evident. The first spans the Scientific Revolution and the second the Industrial Revolution. The growth in the first phase begins to drop off

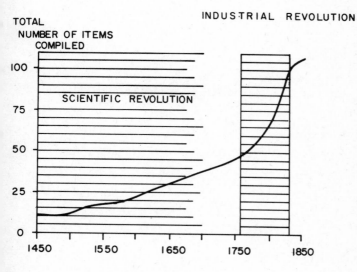

Fig. 3.2   Accumulation of geomagnetic literature in the Scientific and Industrial Revolutions.

at the end of the seventeenth century, and a new more rapid
phase of growth becomes dominant in the second half of the
following century. The growth thus reflects the major time-
divisions in the history of science. The earlier period
of growth appears to be divided into two parts. There is
an initial growth centred on 1525, and a second, longer
period of growth centred on 1675. These may represent
respectively an initial surge of experimentation and simple
measurements following the great explorations, and a longer
period of more sustained growth which occurred when a
larger body of science was available on which to base the
new science, and when the mapping of the geomagnetic field
began to reach the planetary scale.

3.3 Quantitative aspects of the development of scientific studies of the gravitation and figure of the earth

This subject is documented in a readily available form by
TODHUNTER (1873) who analyzed and summarized in a remarkably
thorough manner all of the work in this field from the beginning of the Scientific Revolution to the date of publication
of his work. Thus because of the availability of these data,
this subject is an admirable test case for the quantitative
approach.

The index chosen is the number of published pages

per year. These are averaged over half centuries prior
to 1800 and over decades after. The number of published
works might have, alternatively, been chosen as the index.
However, in view of the fact that the decisive works in
this area of science during the Scientific Revolution were
single-authored books originating from a small number of
highly productive scientists (rather than, as became important later in the history of science, a large number of
shorter contributions by many people), this index appears
to be more suitable. Examples of single authored, decisive
and comprehensive works are the Principia of Newton and
de Horologium Oscillatorium of Huygens, comprising a fair
part of the total in their time period. Because of the
thoroughness of Todhunter's compilation, a unique opportunity
for using such an index is available.

This index, as it varies throughout our period, and
beyond into the nineteenth century is shown in Figure 3.3.
TODHUNTER's (1873) study ends about 1825. Another comprehensive source of data exists spanning the nineteenth
century and was used to carry the study beyond 1825.

This is the article on "Attraction and theory of
potential" in the Royal Society Index (ROYAL SOCIETY, 1909,
pp. 101-109), in which all known periodical literature on
the subject published during the nineteenth century is
listed. These two sets of data are not entirely compatible,

principally because they are compiled in different ways.
The Royal Society publication does not give page numbers,
and also it does not include books or monographs. Thus
the two sets of data are distinguished on the Figure.
However, it is possible to suggest a connection between
them. Their coverage of the field is similar in topics
treated, and a connection may be suggested by comparing
them in the years 1800-1825, when the two sets of data
overlap. In fact a simple conversion, described in Appendix
I, appears to accomplish this connection satisfactorily. The
data from 1800 to 1900 were converted in this way, bringing
them into agreement with Todhunter's data in the time-
period of overlap. In the Appendix it is shown that this
simple procedure provides a valid linkage of the two sets
of data and therefore that the index for the final 75 years
in Figure 3.3 is a valid continuation of that for the
earlier years.

A sequence of peaks in the rate of accumulation,
representing a sequence of growth curves in escalation
when the cumulative total is considered, similar to what
we found for geographical discovery, is apparent in the
Figure (3.3). The points of maximum growth are centred
on 1690, 1750, 1770, and 1825. Following these four phases
of growth is a long stretch of steady growth extending beyond
the final date portrayed in the Figure.

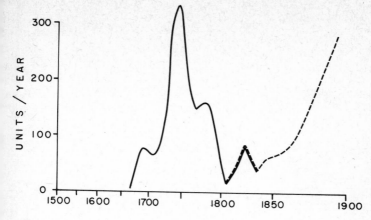

Fig. 3.3   Rate of accumulation of literature on
           mathematical theories of earth's shape
           and gravity.   The units employed are
           explained in Sections.

3.4   A quantitative study of the development of geological
sciences

A documentation of this subject through the Scientific and

Industrial Revolutions is available in von Zittel's "History

of Geology and Paleontology" (1901).   This is regarded as

a standard work for the period covered and gives a compre-

hensive discussion of published literature on the subject.

The index for this study was chosen as the number of published

works per decade, and is shown on Figure 3.4.   Such an index

NUMBER OF
PUBLICATIONS PER DECADE

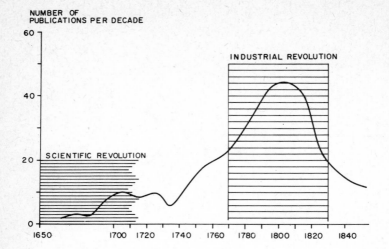

Fig. 3.4   Rate of accumulation of literature
            in the forerunners of the geological
            sciences.

measures the rate of accumulation of geological literature
(Sec. 1.1). The curve shows two peaks, the first beginning
about 1700, evidently representing a growth phase beginning
in the Scientific Revolution. The rate of growth flattens
off during the first decades of the century. It is superseded
in mid-century by a vigorous growth, which sweeps upward
to reach a peak in rate of growth about 1800. In terms of
growth of the subject, this second phase of growth began
at the middle of the seventeenth century, passed its
point of most rapid growth about 1800 and reached its

upper level in the middle of the nineteenth century. This phase of growth coincides with the Industrial Revolution. It represents the first period of strong growth of the geological sciences, and the connection of this growth with the Industrial Revolution is confirmed in a later chapter (Sec. 4.2) by a quite different approach. In that section the origins of the ideas of James Hutton, a key figure in the development of the geological sciences, are analyzed. This growth phase ends with the first modern system of geology, developed by Charles Lyell and expounded in his Principles of Geology (1875). Even if a powerful social stimulus comparable to that which appeared during the Industrial Revolution had acted on the geological sciences during the Scientific Revolution, extensive growth of these sciences may not have occurred because at that time such sciences as thermodynamics and chemistry, required for the advance of the geological sciences, were not sufficiently developed. It was during the Industrial Revolution, when these sciences had developed and when powerful social stimuli appeared, that the critical phase of growth of the geological sciences occurred.

3.5 Interrelation of the growth curves

Each of the subjects: geographical exploration, geomagnetism, gravity and figure of the earth, and the geological sciences shows a peak in its rate of accumulation in the period of the Scientific Revolution, indicating for all of them an episode of growth spanning the period. A comparison of the growth curves may be made by normalizing the cumulative total of each to 100% at the end of the Scientific Revolution. Figure 3.5 shows the time-relations among the growth curves. The growth rate of geographical exploration was maximum for this period at a time centred on 1560 (see also Table 3.1). We have noted on a number of occasions that the expansion of overseas trade, navigation and exploration was the principal factor in the upsurge of scientific

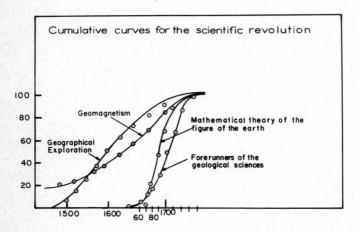

Fig. 3.5   Accumulation of several of the earth sciences during the Scientific Revolution.

interest in geomagnetism and in the gravitation and figure of the earth, which occurred during the Scientific Revolution. This sequence can be seen in Figure 3.5. It took a little more than one hundred years, until 1675, for the maximum rate of growth of geomagnetism to occur, and a somewhat longer lag was required, until 1690, for the maximum rate of growth for gravity and figure of the earth to occur. There was an earlier burst of growth of geomagnetic studies in the Scientific Revolution, concurrent with the first rise in geographical exploration. This was the phase in which determinations of declination and inclination, made in conjunction with overseas voyages, were spreading over the surface of the earth. The maximum rise, centred on 1675, occurred in the period when accumulation of these observations was leading towards world-wide maps of them and towards the beginnings of a theory of geomagnetism. This phase of growth represents a movement in science which began with the great navigations. The new science of geomagnetism would not have been possible unless basic scientific advances in magnetism and in related techniques such as cartography had developed. The time lag represents the period necessary for these advances to take place and to be absorbed and applied. Details of these develop- ments have already been discussed (Sec. 2.5). The study of the gravitation and figure of the earth required a

similar but longer time lag (until 1690) for maximum increase
in this particular phase of its growth. The amount of
basic science required for advance to be made in that
subject was much more than was necessary for geomagnetism
to develop.

3.6 The shift of initiative

Gravity and figure of the earth

The two main streams of development of the theory of this
subject were in France and England, with the final formulation of basic theory taking place in England. Immediately
following the Scientific Revolution, the course of development of maritime activity and exploration was somewhat
different in France than in England. Maritime voyages were
dropping off in other countries early in the eighteenth
century but rose in France, reaching a high point by mid-
century. Reflecting this growth is the high peak in the
curve for gravity and figure of the earth (1750), attesting
to a great wave of scientific work on the subject, hardly
duplicated in its rate of production until a long time after.
Thus we see an example of an important causative chain which has
lain at the root of earth science through later periods
as well as during the Scientific Revolution. Overseas
navigation led to the stimulation of geographical discovery

followed by the growth of sciences based on them. The movements through time and space of geographical discovery and the sciences of the gravity and figure of the earth are shown in Figures 3.7 and 3.8.

A third phase of growth (Figure 3.6) is centred in France following a shift of initiative in the subject from Britain to France, to be explained in the following sections. The peak in the decade 1820-30 coincides with a surge of worldwide exploration of the earth's gravity.

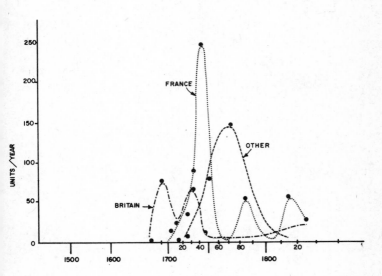

Fig. 3.6   Rate of accumulation of literature
            on mathematical theories of the earth's
            shape and gravity, selected countries to
            1820.

190

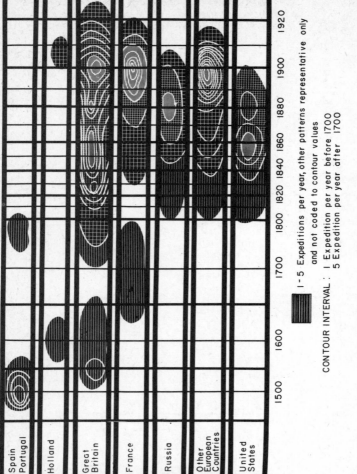

CONTOUR INTERVAL : 1 Expedition per year before 1700
5 Expedition per year after 1700

1-5 Expeditions per year, other patterns representative only
and not coded to contour values

Fig. 3.7   The shift of initiative in geographical
discovery. The data in Fig. 3.1 are
displayed here in both time and space.

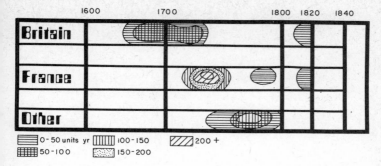

Fig. 3.8   The shift of initiative in mathematical
            studies of the earth's shape and gravity
            to 1820. The data in Fig. 3.6 are dis-
            played here in both time and space.

field, carried out by many nations. LENZEN and MULTHAUF
(1965) detail some of these expeditions.

    The prominent episode of growth in countries
other than Britain or France, as seen on Figure 3.6,
centred on 1770 was caused largely by the upsurge of mapping
and associated geodetic operations on the continent of Europe.
As pointed out by SKELTON (1958, p. 610) these national
surveys "sprang from the same impulses as their prototypes
in France and England...In an age of almost continuous warfare,
governments appreciated the necessity for accurate maps in
political and military intelligence."

    We have asserted in the opening chapter that the
development of sciences is related to development in

general history. The following instance will serve to illustrate the relationship of general history to the growth outlined in the present chapter. The shift of initiative in mathematical theories of the figure of the earth from England to France which occurred between 1700 and 1730 is related to the histories of the two countries in a way which makes it an excellent example.

Science in Britain and France

(a) The early eighteenth century pause. Science declined seriously in almost all of Europe except France in the first part of the eighteenth century (Sec. 2.7).

In England, the pace of science had risen swiftly in the last half century of the Scientific Revolution. The particular timing of this quickening of science has been attributed to the establishment of the Commonwealth at the close of the Civil War, in 1645, in the context of the mercantile expansion of the period. At this time the feudal aristocracy was replaced by the rising merchant capitalists as the prime holders of political power. In this context, a scientific movement primarily caused by mercantile expansion was further encouraged and stimulated.

The Restoration (1660), and the partial return of the feudal aristocrats to power began to change the prospects

of science. The return to a state where science was controlled only indirectly by the forces of mercantile expansion (and in any case this expansion was waning as the
seventeenth century drew to a close) meant a visible decline
of science in England.

In contrast to the general experience in Europe,
the early eighteenth century saw a rise in scientific
activity in France. This occurred for a number of reasons:
it stemmed partly from court circles, and partly from a
rising middle class, as an expression of dissatisfaction
with affairs. This dissatisfaction was to continue until
the French Revolution of 1789 (BERNAL, 1965, p. 361).
France continued to have a strong interest in overseas
trade, although suffering a sharp setback when she lost
her North American colonies in 1763. Even though the feudal
aristocracy was not overthrown until 1789, France had
taken part in the great trading ventures of the preceding
centuries, and an influential class of merchant capitalists
had developed in that country. For these reasons many of
the developments of the Scientific Revolution had taken
place in France. An important part of this development
was the founding of the Académie Royale in 1666. The
Académie was to have great influence upon the earth
sciences during the following years and especially during
the early eighteenth century. In this period France ruled

a wide empire and in that country science in the interest
of trade continued to grow. This growth continued until
the loss of her colonial empire in the mid-eighteenth
century. It was in this setting that the study of the
gravity and figure of the earth moved decisively to France.

3.7 The fruits of these developments: a new earth science

For reasons already explained (Sec. 2.6) France sent out
expeditions to the equator and the Arctic circle to determine
the shape of the earth. The heroic labours of the
members of these expeditions lasted from 1735 until 1744.
In the course of these researches new possibilities in
the study of the earth's gravitation emerged.

The main aim of the expeditions was to measure
the arc length on the earth's surface corresponding to one
degree of latitude. This quantity varies from equator
to pole (Fig. 3.9) in a manner related to the earth's shape,
and therefore was of prime interest to the expeditions.
However, they also made pendulum observations, not surprisingl
since the period of a pendulum is also closely tied to
the shape of the earth (Sec. 2.6). The equatorial party
worked in one of the great mountain areas of the earth,
in the Andes. These conditions engendered special problems
and additional difficulties for the interpretation of the

195

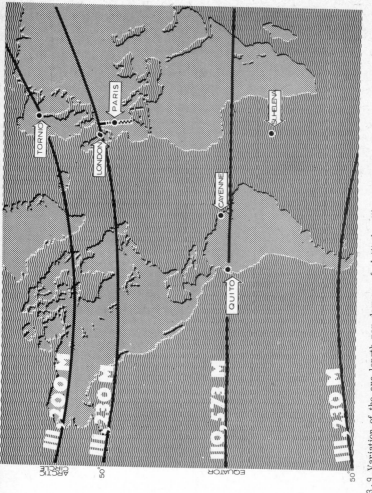

Fig. 3.9 Variation of the arc length per degree of latitude with position from equator to poles. The restricted length of the French arc is shown in contrast to the distance between the scenes of operation of the polar and equatorial parties.

pendulum results. In the course of interpreting such results Pierre Bouguer found that attempts to correct for the effects of the terrain led also to a method of calculating the density of the earth's deep interior. This rather surprising connection between the earth's most elevated parts and its deep interior advanced the understanding of the earth to a considerable degree. It is interesting to note that the importance of measurements of the vertical variation of gravity had been recognized in an earlier period, particularly by Robert Hooke (Sec. 2.6).

In this development we discern an approach to science extending from Francis Bacon through Robert Hooke, then coming to fruit with Pierre Bouguer. One aspect of the approach of these scientists was that they were primarily experimentalists as opposed to pure theoreticians. This difference is exemplified in the contrast between Isaac Newton and Robert Hooke. A significant link among these three figures in science is their exceptionally high regard for experiment. This view of experiment is very much a part of the Baconian approach. Bacon, in his outline of a general programme for the development of science, stated that real advance in science requires the facts of deliberate experiment" (Sec. 1.5). Regarding Robert Hooke, CROWTHER (1960b , p. 222) says that he "made a major contribution to the general programme for the general development of science which Bacon had delineated. He possessed in the highest degree the experimental skill... necessary to make the Baconian programme fertile". And finally, of Bouguer, MIDDLETON (1961, p. xi) remarks that "Bouguer was first and foremost an experimenter, in the tradition of Boyle and Hooke".

As for the experimental versus theoretical approach to science, the frequent collision of the two even in the Scientific Revolution (when often experimenter and theoretician was one and the same person), is exemplified by the controversies between Hooke and Newton. For various reasons the theoretical approach became overrated in England and some authors ascribe the decline of science in England after 1700 partly to this cause. CROWTHER (1960b, p. 222) remarks that "Newton's prodigious achievement caused mathematics and theory, against his own principles, to be temporarily overrated, and the universal experimentalism of Hooke to be underrated." As a consequence many of the problems posed by Newton in Principia Mathematica and the methods applied by him generated a whole chain of research stretching more or less continuously down to the present day. Refinements of problems set out in the Principia Mathematics remained an important part of gravitational research until the early nineteenth century.

Because of the navigational background of many of

the scientific studies of the earth (in which mathematical
formulations play an important part), we frequently find
that problems regarding the gravitation, figure and rotation
of the earth were set out in mathematical terms. Emphasis
was on the capability for calculation of positions, sailing
tracks, map projections and so on. This influence coupled
with the great influence and prestige of Newton assured
the dominance of the mathematical approach for some time to
follow.

Bouguer and the experimental approach

In the course of the stay of the equatorial party in
Ecuador, Pierre Bouguer carried out several pendulum
experiments. In one of the experiments he compared the
length of a seconds pendulum (Sec. 2.6 and Fig. 3.10)
at sea level with its values at Quito (1466 toises above
sea level), and on the summit of the mountain Pichincha
(2434 toises above sea level). He found a measurable
diminution of gravitational attraction (hence shortening of
a seconds pendulum) as he went from sea level to Quito
and from Quito to the summit of Pichincha. He examined
all possible factors such as decrease in temperature with
altitude (which would act to shorten the length of a
metallic seconds pendulum) and lessening of the resistance
of the air to a swinging pendulum at higher altitude,

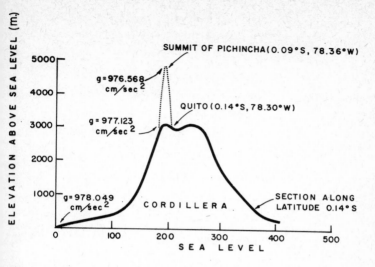

Fig. 3.10   The vertical variation of gravitational
acceleration above sea level, as encountered by Bouguer in his pendulum experiments.
Note that the gravitational effect of
moving through 5000 m. altitude is equiva-
lent to that of moving through about
20 degrees of latitude in low latitudes.

which might affect the results. The effect of latitude
would also be negligible since all three points of observation were at approximately the same latitude. Bouguer
therefore concluded that the diminution of of gravity measured
when proceeding from sea level to Quito and then to the top
of Pichincha could be caused only by the fact that at higher
altitude the observer is farther from the centre of gravity
of the earth as a whole, hence subject to a smaller gravi-

tational force. This effect is offset by the downward
attraction of the matter comprising the elevated plateau
or mountain upon which the observer is stationed, by virtue
of the fact that it lies below him. The first effect acts
in proportion to the mean density of the earth and the second
is proportional to the mean density of the elevated topo-
graphy. Bouguer was able, by considering the actual elevations
involved, to calculate the ratio of densities. His results
made it immediately clear that "the Cordilleras of Peru in
spite of all the minerals they contain have less than 1/4
the density of the interior of the earth...thus it is
necessary to admit that the earth is much more compact...in the
interior than at the surface...Those physicists who imagined
a great void in the middle of the earth, and who would have
us walk on a kind of very thin crust, can think so no longer.
We can make nearly the same objection to Woodward's theory
of great masses of water in the interior" (Bouguer 1749,
quoted by MACKENZIE 1865).

Bouguer was somewhat dissatisfied with the actual
values of the density ratios that he had obtained, although
they left him in no doubt regarding the general conclusions
quoted above. Newton had shown from astronomical considerations that the earth's mean density is between 5 and
6 gm/cm^3. Knowing that and taking into account the fact
that the "soil of Quito is like that of all other countries"

and therefore with a density between 2.5 and 3 gm/cm^3, Bouguer knew that the ratio of 1:4 between Cordillera and whole-earth densities obtained from the experiments was clearly too low. He and La Condamine repeated the experiment, but in a different form using the effect of the mountain Chimborazo on the plumb line. The density ratio obtained was approximately the same. In explanation Bouguer suggested that since Chimborazo appears to be an extinct volcano it could possibly contain great cavities, lowering its average density. At this point, for reasons to be summarized later (Sec. 3.7), we may mark the beginning of the period in which the study of the earth's shape and gravity can be considered a true earth science.

3.8 The geological sciences

The geological sciences on the other hand grew from the techniques of finding and developing mineral raw materials, from the need to understand soils in agriculture, and from the needs of construction and engineering. They also grew from curiosity and ideological conflict regarding the origin and nature of the earth; but is is significant that these sciences did not enter their principal growth phase until the Industrial Revolution, when the social factors behind them became predominant.

The history of the geological sciences is much
more thoroughly studied than that of the other earth
sciences treated in the present book, and there is a
considerable body of literature on them. The place of
the geological sciences in Figures 3.4, 3.5, and 4.1
as well as in Table 3.2 was determined from data drawn
from these sources.

3.9 A theory of the development of the earth sciences in Europe through the Scientific Revolution

Shown their position in time in Table 3.2 are principal
developments in the European part of the history of three
of the earth sciences: geomagnetism, the earth's shape
and gravity, and the geological sciences.

The main principles of growth evident in the Table
are the following. First of all, in each of the three earth
sciences shown, a definite book or small group of books can
be found which expound the fundamental scientific principles
which the science draws on as its basis. Thus a preceding
phase of scientific development, leading to the basic books,
occurred. Once this critical point is defined as one in
a whole chain of development, it then becomes clear that the
sciences in question do not merely originate in the "great
books", but stretch back, even though sometimes only tenuously
and sporadically, through their forerunners to origins which

TABLE 3.2

Principal developments in the European part of
the history of three of the earth sciences: geomagnetism, the earth's shape and gravity, and
the geological sciences.

Field	First Fore-Runners	Basic Books	Characterizing Work(s)
Geomag-netism	Scientific studies of compass and mag-netism: 1190, 1269 geography-carto-graphy	Gilbert, de Magnete (1600)	Halley; map of declination papers on the variation of declination and the origin of the geomagnetic field 1683-1700
Earth's Shape and Gravity	earth's radius measured from ancient times; pendulum (1583)	Huygens, Horologium oscilla-torium (1687) Newton, Principia (1687)	Bouguer's experiments in Peru (1737-8)
Geological Sciences	Agricola de re Metallica (1550±)	Hutton, Theory of the Earth (1795)	Lyell, Principles of Geology (1830)

were centuries or in some cases millennia in the past. These forerunners usually have contributed much to the important basic work, although the latter may have drawn also on other sources. Because these basic books are related to the principles behind the particular scientific field, they lie in its "forerunner" stage of development. Following the "basic scientific book" we find some experiment, programme of observations or publication which mark the beginning of the subject in its true form. These relationships are shown in Table 3.2.

The point at which a discipline in earth science comes into being is when its forerunners become spread over the earth's surface, or embody experiments which are conducted through a sufficient time-span, that the scale of measurement begins to match the scale of the earth phenomena involved. Several lines of development in science, some of them originating in the very early civilizations, reached the required scale during the early phases of modern science in Europe. Since they were related to important social concerns regarding the earth, and to important scientific aspects of the earth, they developed into branches of earth science.

The birth of geomagnetism

The forerunners of the science of geomagnetism follow a

direct chain of theory and experiment dating back to the
invention of the compass and measurements of the declination
in China. This line of development, as it moved from China
to Europe and then through the technical developments of
the late Middle Ages in Europe to grow rapidly under the
stimulus of the new ideas of the Renaissance and finally
flower during the closing years of the Scientific Revolution,
was traced in an earlier chapter (Sec. 2.5). The basic
scientific theory which made it possible for geomagnetism
to emerge as a science was developed in the late Middle
Ages and in the middle phase of the Scientific Revolution.
This science was brought into a sufficiently powerful body
of theory by William Gilbert in de Magnete. The same forces
which impelled the development of the basic scientific
background sent magnetic measurements across the globe and
stimulated sufficiently sustained interest in these measurements, that Edmund Halley had, 83 years later, data on the
earth-scale spread through space and time. These data
allowed him to construct world maps of declination and to
arrive at a fruitful theory of the origin of the geomagnetic
field and the cause of its variation in time. Thus the
science of geomagnetism, as distinct from its forerunners,
may be dated from 1683, when Halley's first paper based
on these data appeared (Sec. 2.5).

The birth of the science of the gravity and the figure of the earth

The fcrerunners of this science likewise are parts of a
chain originating in the early civilizations. Geodetic operation
allowing the calculation of the earth's radius were carried
out in the Mediterranean area in Alexandrian times, in
China during the T'ang dynasty, and in the Arab world in
the Middle Ages. As with geomagnetism, the upsurge of the
Scientific Revolution created basic theory and sufficiently
accurate instruments to make the development of the new
science possible (Sec. 2.6). The subject was founded on two
basic books: Newton's Principia and Huygens' Horologium
Oscillatorium. The French expeditions of 1735-44 mark the
first spread of co-ordinated geodetic operations on the earth-
scale. One part of these operations is chosen as marking the
birth of the new science, the series of gravitational measurements begun by Pierre Bouguer in 1737 (Sec. 2.6). This point
is chosen because from that time on experiments conducted at
the surface were used to infer conditons in the earth's interior, thus introducing another dimension on the earth-scale
to the study of the earth. It might be argued that the point
of departure for the new science should be some years earlier,
starting from Newton's correct interpretation of Richer's
pendulum results (which also extended observation on the
earth-scale). However, this interpretation was part of

what was then a theory regarding the shape and gravitation of the earth, the first dramatic proof of which came with the French expeditions of 1735-44. Some people might even argue that Newton's _Principia_ cannot be regarded as one of the basic works preceding the establishment of the new science if we choose to mark its beginning by the French expeditions because the expeditions started out firmly in the belief that Newton was wrong. However, most of the expedition members became convinced Newtonians after seeing their results confirm Newton's predictions (Sec. 2.6). Thus on both these counts, the interpretation offered above appears to be the most valid. It should be remembered too that although the science of the earth's shape and gravitation came into being after the close of the Scientific Revolution had for the most part ended, it is nevertheless primarily connected with the Scientific Revolution by virtue of its origins in France where special conditions extended the thrust of this period for some decades (Sec. 3.5).

Chapter 4

THE INDUSTRIAL REVOLUTION (1760-1830)

We have now seen how a number of branches of earth science
have come into being. The development at these points
of origin of earth science is evidently of a type which
has continued through a long span of history (Sec. 1.3).
Having established the nature of this development, let us
now see whether or not it is a reliable guide into the period
of the Industrial Revolution.

4.1 The Industrial Revolution and science

The great changes in Europe which had such profound
effects on science during the Industrial Revolution had,
besides leading to growth in the volume of trade and
manufacture, led to a continued and cumulative change
in their character. An economic setting in which technological change was to set off a period of unprecedented
economic growth had developed by the end of the seventeenth
century (COLE and DEANE, 1965, p. 3). Cities had grown
rapidly, many large-scale industries had developed,
specialization of labour had begun, and elaborate overseas
trading connections had been developed. The technical pro-

gress made during the Scientific Revolution had led to a
wealth of invention. These new methods proved to have a
potentiality in manufacture many times greater than that of
the older methods. Regions where iron deposits and coalfields
were in abundance and close together were suddenly elevated
to a position of potential leadership in industry. Northern
England and Scotland particularly possessed all the necessary
factors in the most favourable combination. As a result
there occurred in those regions a new movement in manufacture
which soon became a headlong transformation so rapid and
thorough as to become known as a revolution, the Industrial
Revolution. A sharp rise in productivity and production took
place, and was well under way by 1760. In this, and in subsequent decades, new factories and new factory towns sprang
up. An irreversible change to a new age based on coal, iron
and steam occurred. This revolution, begun in Britain, spread
to other European countries beginning with France, Belgium
and the Netherlands. It began to be felt overseas, particularly
in the United States, after the turn of the century. The
change was virtually completed in Britain by 1830, although
it did not reach its full momentum in other countries until
mid-century.

Even though the Industrial Revolution itself did
not at first have its full effect upon manufacturing in
countries other than Britain until some decades after

1830, ideological and political counterparts appeared in these countries at the same time as the industrial upsurge in Britain. These counterparts came into being in the following way. The Scientific Revolution and the developments which generated it left some mark on all countries which experienced them. For one thing, there was a widespread desire to break with old ways. Part of the vision of the future came, for a group including many intellectuals, from an awareness of the possibilities of improvement through science and technology. This awareness became an ideal which inspired political and scientific movements even when science-based improvements did not take place. Such a movement when combined with deep dissatisfaction at the lack of change sometimes contributed to the occurrence of violent upheavals, such as the French Revolution.

A change so radical and so closely bound up with science, as the Industrial Revolution was, could not help but affect science itself. This effect on science in Britain, where the economic transformation first took place, had been traced in detail by CROWTHER (1962). As in other sciences, the earth sciences were led in new directions by this industrial upsurge. The growing metallurgical and chemical industries were based on raw materials extracted from the earth. This emphasis on mineral resources

led directly to a rapid growth of the geological sciences.
In previous periods, when overseas trade was relatively
more important than manufacture, the sciences that were
later to form the geophysical sciences reached a higher stage
of development than those that led to the science of geology.
Because of their lesser social importance in the earlier
period, the geological sciences had never been scientifically
synthesized to the degree that the geophysical sciences had
been. However, the Industrial Revolution while continuing to
stimulate these latter sciences, rapidly accelerated the
former into a comparably important position.

4.2 The geological sciences and the Industrial Revolution

The forerunners of the geological sciences had, before the
Industrial Revolution, passed through a period of growth
(Sec. 2.7). However, their first major growth phase still
lay ahead of them, and this initial period of (small) growth
was cut off in the decade 1730-39 by a new vigorous phase
of growth (Fig. 3.4). This growth, stimulated by the
Industrial Revolution, led to a further development
of the long-standing influence of mining in the Freiberg
district. This was the founding of the Freiberg School
of Mines in 1766. This became one of the leading schools
of geology, forming one of the streams which led to modern

geology. It was at Freiberg that Abraham Werner established
a positional scheme of stratified·rocks and established
the mineralogical composition of rocks as an important
part of their classification. This line of development
was to become the beginning of modern observational geology
(TOMKEIEFF, 1950, p. 389). Alexander von Humboldt, who
was later to have considerable influence upon many branches
of earth science (including geomagnetism) received part
of his scientific training in the School of Mines, 1790-92.

This was the first phase in the development of
the geological sciences in the Industrial Revolution.
It was soon to be followed by even more decisive developments.

James Hutton, founder of modern geology (1726-1797)
The above heading is the title of an address given by
a distinguished British geologist to the Royal Society of
Edinburgh on the occasion of the 150th anniversary of
the death of Hutton (BAILEY, 1950). The correctness of
Hutton's geological conceptions and of his theory of the
earth is attested to by the high place he continues to
hold in science. Hutton's principal theories of geology
were first set forth in a paper read to the Royal Society
of Edinburgh in 1785 and appeared at greater length in
his book, Theory of the Earth published in 1795. Some

elements of his theory can be traced back to ancient
times (BAILEY, 1967, p. 45), and through the decades
preceding Hutton's work, many attempts had been made to
formulate theories of the earth. Several branches of
what were to become geophysical sciences, as well as
crystallography had been established. Yet no system of geology
comparable to Hutton's appeared earlier.

It is now recognized that the time and place of
the appearance of the first valid system of geology (that
of Hutton) were not accidental (BAILEY, 1950, 1967;
TOMKEIEFF, 1950; CROWTHER, 1962; McINTYRE, 1963). The
above mentioned authors present a considerable body of
evidence that the Industrial Revolution was among the major
influences which 'inspired and required' the theory. The
accelerated development during the Industrial Revolution
inspired an upsurge in science, philosophy, and the
industrial arts. The new industrial centres were in the
north of England and in Scotland. In the last three
decades of the eighteenth century the population of
Birmingham grew fourfold, while that of Manchester and
Glasgow trebled. Centres of science shifted from London
and Cambridge to Manchester, Glasgow, Edinburgh, and
Birmingham. In Edinburgh there appeared a most remarkable
group of scientists and philosophers, which were to become
leading figures in their fields. Among them were Joseph

Black, Adam Smith, David Hume and James Boswell. Animated discussion of science and philosophy continually went on in numerous clubs and convivial meetings. It was in this atmosphere that Hutton found himself while working on his theory of the earth (GEIKIE, 1897; MACGREGOR, 1950, p. 351-6; CROWTHER, 1962).

The Industrial Revolution and Hutton's Theory of the Earth

The relationship between Hutton's ideas and the industrial revolution is a very clear case of the influence of society on scientific ideas. It is therefore important to consider his ideas in their social context.

Hutton's writings and the origins of his ideas have been analyzed at length during the nearly two centuries which have elapsed since the appearance of his book. Some of the main features of his theory are as follows:

(1) The history of the earth's crust is explained by processes that are going on at the present time.

(2) A process which is basic in this history is weathering and denudation, which wear down the surface of the land, transport the resulting sediments, and through rivers carry them to the seas, where they are deposited at depth.

(3) Deep in the seas, these sediments are transformed

into rock and then elevated, forming land again. An internal store of heat in the earth is the agent in the transformation of the materials and their subsequent elevation.

(4) The processes of denudation, consolidation, and uplift are the successive stages in an ever-repeated cycle of wastage and renewal of the land. These processes in the evolution of the earth, which is described as a "machine", are held as evidence of "no vestige of a beginning - no prospect of an end".

Hutton went further and used these principles to predict the occurrence of certain geological features. The idea of cycles of denudation, deposition, and uplift suggested that the "ruins of an earlier world lay beneath the secondary strata" (GEIKIE, 1897, p. 161). The discovery of such an unconformity as predicted by theory by Hutton in the Lammermuir Hills was a "vivid and successful" proof of his theory. The idea of a store of heat deep within the earth suggested that bodies of molten rock might have arisen from below. His demonstration that bodies of granite had indeed intruded overlying rocks in Glen Tilt was another striking proof of the theory. Such conclusions, and the principles behind them are so much a part of modern geology, that it is difficult to realize their novelty in Hutton's time.

Life of Hutton

James Hutton was born in Edinburgh in 1726, the son of
a highly respected merchant who had been City Treasurer.
His father died when he was very young, and James was
brought up by his mother. He went to school and university
in Edinburgh; his higher education was extremely varied,
as was his early life after graduation. He began the
study of the humanities in the university. However, he
soon acquired an interest in chemistry, and he and his
friend John Davie performed many original experiments
together during this period. In spite of his scientific
interests, Hutton entered an apprenticeship in law after
graduation but gave this up after a short period and began
the study of medicine at Edinburgh, Paris, and finally
Leyden, where he received his M.D. in 1749. His thesis was
on the circulation of the blood. On his return to Britain
he did not return to Scotland but settled in London.
However, he did not practice medicine. Due to contact
that he made with his old friend, John Davie, he became
interested in a scheme to turn one of the experimental
researches which he and his friend had carried out long
before (on the manufacture of sal ammmoniac out of soot)
into a commercial venture. This enterprise brought him
back to Edinburgh. Shortly after, he inherited a farm
close by and decided to devote his interest to farming,

and soon became immersed in the modern notions of mechanized
farming. Scientific agriculture in that period was of as
great significance as the Industrial Revolution, and indeed
part of it. In 1765, he moved back to Edinburgh and
formally joined Davie in the sal ammoniac business. He
continued all the time to make geological excursions
and continue his writing on geological and philosophical
topics. Thereafter, he remained in Edinburgh amid the
stimulating surroundings and his circle of scientific
and philosophic friends. It was during this period
that he wrote his Theory of the Earth.

Sources of Hutton's ideas

The principal ideas behind Hutton's system are as
follows:

(1) Basic to everything is the principle (later
to be known as "uniformitarianism") that "the present is
the key to the past". Bound up with this principle is
Hutton's method of proceeding to conclusions by an induc-
tive process from observations made about the earth.
This was in contrast to the attempts to move from general
principles towards conclusions about the earth, so common
in previous theories.

(2) Of equal importance is the idea that
subterranean heat is the primary source of energy for

the evolution of the earth; this heat is held to be
capable of elevating strata from deep in the seas to
form elevated land areas, as well as of consolidating
sediments in the sea into rock.

(3) Following this idea is the conclusion that
the earth must evolve by means of great cycles of denudation, consolidation, and uplift.

(4) These last two ideas in turn led to the
conclusion that igneous bodies that have risen from
below and unconformities must be important elements in
the geologic section.

(5) Hutton's view of the earth was that it is
a machine constructed on chemical and mechanical principles,
with heat as its motive power; but at the same time it is
also an organism in which the processes of decay and repair
constantly go on "in a salutory circulation".

(6) Chemistry was an important part of Hutton's
system; he applied contemporary discoveries in chemistry
to develop certain parts of his theory. He viewed the
chemical constitution of the earth as being very important.

These ideas were all ones which can be identified
with the Industrial Revolution. The idea that the earth
is a heat engine, with heat as the motive force in cyclic
action, is certainly one which must have been compelling
in a world being shaped by the new steam engine invented

by James Watt (who was one of Hutton's friends). The comparison with an organism, with a continual circulation of matter has been ascribed to Hutton's studies in medicine (TOMKEIEFF, 1950; McINTYRE, 1963). Also, Hutton's years as a scientific farmer should not be forgotten in this context. The fact that Hutton considered the chemical constitution of the earth, and uses chemical reasoning to establish some of the basic processes have been traced to the influence of Joseph Black (a pioneer of modern chemistry, who also was one of Hutton's friends).

We have listed several features of Hutton's system: heat as basic force, leading to the elevation of strata and the consolidation of sediments; the earth as a machine and a chemical system; uniformitarianism; the earth as an organism; and cycles of decay and repair. The ideas related to heat and chemistry, picturing the earth as a machine are characteristic of the Industrial Revolution. The development of these subjects in this period was influenced strongly by Hutton's friends, Black and Watt. The remaining three ideas show traces of different origins.

Uniformitarianism is closely allied to the scientific method of Francis Bacon (CROWTHER, 1962, p. 82). The remaining two, as already mentioned, have been traced to Hutton's studies in medicine. One aspect of Hutton's

view of the earth as an organism is related to the
Industrial Revolution in another way, however. A distinguished student of Hutton, E. B. BAILEY (1950, p. 359)
remarks "I have always felt when reading his geological
publications that he wrote of the earth as a well-managed
agricultural estate with a rotation designed to maintain
continuing fertility." This connection between Hutton's
geological ideas and farming is significant. It was in
the course of his contact with nature as a farmer, and
particularly in the course of his deep thinking about
nature from the point of view of a modern mechanized
farmer that Hutton became interested in geology. Hutton
wrote a treatise on agriculture which he considered as
his principal work. In Hutton's time a revolution in
agriculture, as profound as that in industry (and
connected with it), occurred. Concentration of population
in the new industrial cities, and the growth in population
which accompanied the Industrial Revolution resulted in
"the new cash-crop agriculture of the landlords and farmers,
who were replacing the peasants and their subsistence
agriculture over most of England" (BERNAL, 1965, p. 369).
This new agriculture, connected with the Industrial
Revolution, was imbued with a similar spirit of innovation
and progress. As BERNAL (1965, p. 369) remarks, "the
agricultural revolution was a mixture of empirical breeding

and crop rotation and mechanization".

James Hutton was an extremely versatile man, acknowledged by those who knew him, as well as by his biographers as a man of genius. We have seen that it was the course of his life, and his surroundings, as determined by the Industrial Revolution that brought Hutton to the study of geology. Many of the observations used by Hutton to establish his theory, and many of the basic ideas of his system had been known before his time. But the correct combination of ideas, and the principles needed to guide thought correctly through the many complex possibilities for the operation of the earth, could not have occurred before the Industrial Revolution. Thus it was that the geological sciences were not established as such before this period.

4.3 The Enlightenment and the earth sciences

James Hutton is an example of a scientist who lived amid and was part of an economic transformation of the first order. We have seen that this situation was a principal factor leading to Hutton's achievements as "the founder of modern geology". In other parts of Europe the situation was different. An economic transformation comparable to the Industrial Revolution in Britain did not occur on the

continent of Europe until much later. Thus we might expect
to see a different sort of influence on science occurring
in countries in this geographic area. However, an intellectual ferment (the Enlightenment) did occur on the
continent contemperaneously with that period. This upsurge
of ideas had its effect on the earth sciences as illustrated
by the life of Alexander von Humboldt, who contributed in
his explorations and scientific work to geology, geophysics
and geography. The similarities and differences between
Hutton and von Humboldt are significant.

Alexander Von Humboldt (1769-1859)

The career of von Humboldt was quite different from that of
Hutton, although the scientific contributions of both to
earth science were of comparable importance. Von Humboldt
was born in Berlin, at that time the capital of the Prussia
of Frederick the Great, in 1769. Like Hutton, von Humboldt
was born into a well-to-do family, and like Hutton, his
father died when he was very young and he was raised by
his mother. He was descended on his mother's side from
a family of French Huguenots, who had founded a plate-glass
factory in Brandenburg. This background is remarkably
similar to that of Dmitry Mendeleyev in Russia who likewise
had a considerable interest in the earth (Sec. 1.6).
However, Mendeleyev's situation was somewhat different

in that industrialization had begun in Russia during his lifetime. During von Humboldt's youth the effects of the Industrial Revolution had hardly begun to spread to Prussia except in a few isolated instances (COLE and DEANE, 1965, p. 15). In countries which did not experience the Industrial Revolution, particularly France, which had a total production comparable to that of England in the mid-eighteenth century, and was therefore potentially an important industrial country, there was a considerable interest in science (COLE and DEANE, 1965, p. 11) and an awareness of what a science-based industry could do. For various reasons, however, industry continued on the old lines. This situation gave rise to a considerable amount of dissatisfaction among scientists and in other circles interested in science, combined with a keen interest in enlightenment and progress. These ideas were to have their fulfilment in the French Revolution and many who shared them showed interest in and sympathy with the Revolution. These ideas spread to other countries, most of which had at least a small circle of people imbued with the progressive spirit of science. Von Humboldt as a young man was introduced to such a circle in Berlin at the house of Moses Mendelssohn (KELLNER, 1963, p. 8). This circle included Marcus Hertz, a disciple of Immanuel Kant, who delivered lectures on philosophy and science.

In addition to the spirit of scientific progress, von Humboldt may have begun as a young man in Berlin to absorb the philosophy of the unity of nature, and an appreciation of the intimate interrelation that was held to exist between electricity, magnetism and living things that was so strong in Germany at the time. This interest can be seen later shaping von Humboldt's scientific work, although at a later period he disagreed strongly with the excesses of German nature-philosophy (KELLNER, 1963, p. 115). During his university education von Humboldt studied a wide variety of subjects gradually moving towards physics and chemistry and particularly natural history. These latter studies developed in him an interest in geology, which brought him in 1790 to the Freiberg School of Mines as a student. His professor was Abraham Werner, one of the founders of economic geology. Werner was probably also influenced by the school of German nature-philosophy (MASON, 1953, p. 323). This influence may have helped to develop the unitary approach to nature which is evident in von Humboldt's later works. At the same, time, von Humboldt became one of the most influential supporters of empirical science in Germany, as opposed to the older nature-philosophy. This view of science was part of the new rising ideology of the industrial age (THOMAS, 1951, p. 28).

After graduation from the School of Mines in 1792

von Humboldt was appointed to the Mining Department of the
Prussian government and soon became inspector of mines in
the Fichtel Mountains. He remained in this post until
1796. He had the broadest interest in science of all
kinds: botany, mineralogy, geology, meteorology, electricity,
magnetism, physiology, chemistry. This unitary approach
to nature was typical of science and philosophy in Germany
at the time (MASON, 1953, pp. 280-290). In this view of
nature all phenomena are considered as a unity and the
forces of electricity and magnetism were regarded with great
interest as a possible moving force in nature including
living things and their evolution. During his stay in the
Mining Department, von Humboldt discovered and mapped an
interesting magnetic anomaly in the Fichtel district (KELLNER,
1963, p. 22). Also at this time he performed experiments
on galvanic effects on living tissue (KELLNER, 1963, p. 19).
This unitary approach was evident in the scientific measurements that von Humboldt made during his famous voyages to
South America and Siberia. During these expeditions he
surveyed a broad range of sciences: geology, flora, fauna,
atmosphere, oceans, anthropology, and archeology. Among
these, geophysical and meteorological phenomena received an
important place in which measurements of significance in
geodesy, geoelectricity, air pressure, air temperature,
chemical constitution of water and air formed a prominent

part. We shall see that because of this comprehensive interest, von Humboldt was to influence one of the leading figures in physics and mathematics and bring him into studies of the earth (Sec. 5.2).

During his lifetime von Humboldt made contributions to many of the earth sciences. In the present book we shall be concerned particularly with his contributions to geophysics. We will come across his influence on geomagnetism a number of times. His own entry into this science was, as we shall see, largely under the influence of the French scientific establishment. This association came about because of Humboldt's growing interest in travelling abroad. Having resolved to take part in a scientific expedition of some sort, he resigned from the Prussian Mining Department in 1797. After several false starts in various places, Humboldt arrived in Paris in the following year. France was at the time supporting many scientific enterprises of the type in which Humboldt was interested. The special position of France in this regard is a matter of interest.

Science in France in the late eighteenth century

As has been mentioned in preceding sections scientists in France, as the monarchy drew to its close, were restless because of the failure of science to reach its full

potential for improvement of industry and other spheres
of life. The revolutionary governments which followed
the French Revolution were equally desirous of seeing
science contribute to the advance of the country. Under
the leadership of these governments considerable state
support for scientific projects was given and a system
of modern scientific education was instituted, the first
in Europe. There grew up a body of scientist-professors
much like those of the present day, recruited from talented
young graduates of the new institutes. These formed an
extraordinary group which was to make significant advances
in many sciences. As a result, France gained scientific
predominance, which it maintained well into the nineteenth
century. The Revolution soon gave way to the Napoleonic
period, during which time the scientific drive continued.
This development was in fact markedly speeded up during
the early phases of the Napoleonic period (BERNAL, 1965, p.
380). As the Empire wore on, however, the pace of science
slackened (KELLNER, 1963, p. 63).

Humboldt in Paris and his first great exploration
Soon after his arrival in France, Humboldt took part in
the geodetic operations that were being conducted throughout
the country, as part of an early mapping programme, one of
the first in Europe. Following this work, Humboldt

was offered a place on an expedition which was to sail to
the Pacific regions and ultimately reach the South Pole.
During the preparation for the expedition, Humboldt
adopted the Chevalier Borda as his principal advisor because, as
von Humboldt remarked, he exerted a "decided influence.....on
those who were equipping themselves for remote expeditions"
(VON HUMBOLDT, 1858, p. 61). Borda interested Humboldt in
geomagnetic measurements and invited him to "choose all the
instruments he might need from the national collections"
(KELLNER, 1963, p. 28). Thus Humboldt was influenced
by the fact that in the decades preceding his arrival in
France, there had been considerable interest in geomagnetism.

The expedition, however, did not sail because of
financial difficulties arising from Napoleon's campaign
in Egypt. Disappointed, Humboldt set out to find another
opportunity to sail on an expedition. He finally obtained
permission from the King of Spain to travel in the Spanish
possessions in South America. Among his scientific equip-
ment was one of Borda's magnetometers. During the course
of his travels, Humboldt made major contributions to geomagnetism. The expedition was entirely paid for by
Humboldt himself, from an inheritance. Even though the
expedition was conducted on Humboldt's own initiative
and paid for by him, its scientific character and objectives
were not independent of the main directions of science, but

grew out of the German and the French streams of scientific development. The extraordinary range of measurements and observations, including atmospheric electricity, meteorology, geomagnetism, geology, geography, botany, zoology, anthropology, and economics show the influence of German science in von Humboldt's day. Humboldt's interest in geomagnetism reflects the upsurge of geomagnetism in France at the time, as is made clear by Humboldt's account of the origins of his interest. He cites voyages by Paul Lamanon (1785-87) and Admiral Paul Rossel (1791-4) as the forerunners of his measurements of the intensity of the geomagnetic field in tropical America, made in 1799-1804. These measurements constituted the first unambiguous proof that the force of the magnetic field increases to the north and south of the magnetic equator. This development added field strength as an important element of the geomagnetic field, along with declination and inclination (as used by Halley, Whiston and Wilcke), to be mapped in the search for understanding of the magnetic properties of the earth. Intensity in fact was to become, in the twentieth century, the most important geomagnetic measurement. The French Revolution and the spirit which preceded it, was a powerful influence upon Humboldt; the scientific developments which shaped his scientific contributions also came from this source. Thus general history determined the development of science at this point in its development, as it always has.

We may now fruitfully compare and contrast
Humboldt with Hutton. Both found their inspiration for
scientific work in the same stream of development. This
was part of the growing understanding in the seventeenth
century that science is an important part of social development. This understanding came from the Scientific Revolution
and found different expression in Britain than it did in
continental Europe. In Britain, the scientific spirit was
expressed in the Industrial Revolution, and it is significant
that Hutton took part in manufacturing and scientific farming,
as well as in science. On the continent, economic conditions
delayed the Industrial Revolution, even though the scientific
world was ready for it. This feeling of frustration led
many intellectuals to support the French Revolution. Thus
Humboldt, in contrast to Hutton, expressed his connections
with society through court and diplomatic activity, sometimes on a very high level. The French Revolution was a
lifelong inspiration to him, leading to his deep feeling
regarding the revolution of 1848 in Berlin (KELLNER, 1963,
pp. 215-16). Thus we see by examining the lives of two
equally important figures in the geosciences, both of whom
were major influences on the development of these sciences,
the contrasting effects of different social environments
on scientists. We see two extremes in the variety of

ways that general history acts on the history of science.

4.4 The main phase in the development of the geological sciences

We have seen by the above two examples how the Industrial Revolution affected the earth sciences and that the origin of the geological sciences took place in this period. The main growth phase in geological sciences spans the years 1730-1830, centred on the decade 1800-1809. This period included the publication of the first geological map (1743), and of the first geological journal (1747) as well as the founding of the first school of mines (1765). This period of growth arose from the increasing need for minerals and from the extension of engineering works (roads, railroads and canals) over the countryside. Various components of the new science sprang up almost simultaneously in a number of European countries. One of the decisive developments have been mentioned: the description of rocks in terms of their mineralogical characteristics (Abraham Werner, in Saxony, Sec. 4.2). Equally important was the correlation of rock units over a large area by William Smith during the years 1790-1815, in which the first use of fossils was made for this purpose. Smith was an engineer engaged in canal cutting and his stratigraphic mapping was done as part of this work. He laid the basis for further extension of his methods by the members of the Geological Society of London (following 1807), leading to the correlation and mapping of Paleozoic formations over much of England. This important advance, as we have seen, came from canal cutting, an operation typical of the Industrial Revolution. The fact that the origins of the geological sciences were found over most of Europe is attested to by the fact that immediately afterwards the concept of fossil-guided stratigraphy shifted to France and was applied to mapping Tertiary formations by Georges Cuvier and A. Brongniart (1808).

This growth phase is prominent on Figure 3.4 as a logistic-type curve, giving the rate of accumulation of geological literature. The peak of the curve occurs at the time of most rapid growth of the subject, and the width of the peak at half maximum may be taken as a measure of the years of greatest growth of the subject. This period spans the fifty years from 1775-1825, and encompasses what has been called the "Heroic Age of Geology" (by von Zittel, 1901, pp. 46-145). In this Age, according to von Zittel, a place was won for geology as a scientific study.

The consolidation of the geological sciences

We have noted that the geological sciences were consolidated during the short period between 1775 and 1825. At the beginning of the period, two of the basic sciences needed for real advances in the subject, thermodynamics and chemistry, were applied to geology by Hutton. Many of the forerunners of the geological sciences similarly found their way into a common stream, as part of the new science. The mineralogical description of rocks, initiated by Werner, and the use of fossils to identify successions of strata over wide areas, as first done by Smith, were two such components of the early phases of geological science. There followed a period of development of the specific organizational forms of the subject: geological maps; systematic field observations fruitfully directed by the new ideas; geological journals and schools; state geological survey organizations; and finally the development of a profession. All of these developments were driven by the Industrial Revolution and its needs.

Table 3.2 places Lyell's _Principles of Geology_ at the end of this period of transformation and as the starting point of modern geological science. Lyell's book merits this position for a number of reasons. The first is that the _Principles_ defined the limits of the subject in a way which has remained acceptable up

to the present time, referring to it as Geology (a term
that had been previously coined in 1778). He defined
geology as the "science which investigates the successive
changes that have taken place in the organic and inorganic
kingdoms of nature; it enquires into the causes of these
changes and the influence which they have exerted in
modifying the surface and external structure of our planet".
Furthermore, Lyell pointed out, the subject had in the
past been "confounded with many other branches of enquiry"
and in particular had been erroneously thought of either
as a subordinate branch of mineralogy or physical geography or as a study closely tied to cosmogony (LYELL,
1875, pp. 1-5). Secondly, and most important, the
Principles firmly established and brought recognition to
one of the important basic ideas of the geological sciences.
This idea suggests that all of the vast changes which occurred
in the past were the result of processes which we see in
operation today. Lyell recognized that vast changes
have occurred in the past, but showed that they could
result from familiar processes provided that the latter
had been operative for a sufficient length of time.
This idea was revolutionary and of great philosophical
importance at the time, and stood in direct opposition to
catastrophism, which had been widely accepted as an explanation of the earth's history. Comparison with our

Section 1.5 suggests a descent of Lyell's ideas from the
Copernican revolution. It is interesting that Bondi, one
of the best known champions of uniformitarianism as applied
to the universe, is at King's College, London, where Lyell
was the first professor of geology.

View of an earth science through history

We have seen, in the preceding sections, the geological
sciences developing into their modern form during the
Industrial Revolution. We have often mentioned the
forerunners of the earth sciences, stating that these
early trends in science went far back in history. The
geological sciences and their predecessors make a good
example of sciences reaching far back in time because a
practical understanding of rocks, minerals and ores came
early in human history. For this reason their extension
into the past will now be considered. Man's earliest
tools were of stone (dating back at least one million
years). Ochres and similar substances were used for
personal decoration and art since Paleolithic times. The
search for suitable raw materials and the experience gained
of the properties of these materials as they were worked
into useful products must have given rise to a great
store of practical knowledge about the earth's surface
and the location and properties of these materials. This

experience and knowledge on the part of prehistoric peoples was the beginning; this and subsequent developments are shown in Figure 3.9. In this Figure we can see lines of interest in the earth stretching back into Neolithic and Paleolithic times. Written records began to appear soon after the ancient and classical civilizations of western Asia, northern Africa and Europe developed. Some of these touched on the earth in its various aspects, and many of the prominent classical writers on science and technology contributed to these forerunners of the geological sciences. With the fall of these civilizations, writing on these topics dropped off. The threads of development were not lost however. Some were carried across to centres of learning in the Arab world while other new threads arose independently in China. With the Scientific Revolution, the principal developments returned to Europe, with a peak of activity occurring during this period (Fig. 4.1). Some of these lines of interest (such as studies of the succession of rocks, fossils, landforms and mountains) were established on a scientific basis during this period, while one of them (mineralogy) assumed some of the characteristics of a modern science. In the following period (the Industrial Revolution), several of these fields were consolidated into their modern forms and most of them came together to form the geological sciences. The Figure is not intended to be a complete record of the development of the geological sciences. A selection of lines of development of contributors to each is shown. The record has not been given in comparable detail in all cases, and no attempt has been made to single out or even include all of the principal contributors. The diagram is intended to show how the forerunner stages of many of these sciences reach far into the past, with the initiative moving from one civilization to another. Short biographical notes on all the contributors included are given in Appendix 2.

4.5 The end of the revolutions in science and industry

These first bursts of development in science and industry in modern times span the period chosen as the primary focus for the present study of earth sciences. The main hope in undertaking the study was that it would show the connections among earth science, science in general, society and history. If such connections can be demonstrated it is possible to go on from this basis towards an understanding of present-day earth science and its future role in human development.

The periods following the Industrial Revolution marked a new era in human society in which it became clear that science and technology might possibly become

238

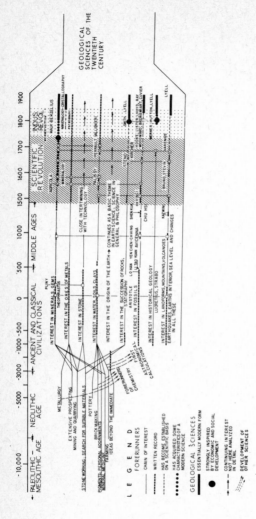

Fig. 4.1   The forerunners of the geological sciences.

determining factors in social development. The full story of how science has moved, during the interval from the Industrial Revolution to the present time, from a position of potential importance to one of crucial importance in history is told by BERNAL (1965, 1971). The period immediately following the Industrial Revolution was a critical one in this development. It was in this period that a sense of progress in development began to be felt (Sec. 5.1), and science was seen to be part of this progress.

It is hoped that the present book will show that in the Scientific and Industrial revolutions there was a definite correlation between social development and that of the earth sciences. These connections had implications for the future; they most certainly apply to the following (the mid-nineteenth century) period. Therefore, the following chapter is included, almost as an addendum, with the hope that it will show that the positive correlations between earth science and society, so apparent earlier, did in fact continue. It is an addendum in one sense; but also it is an introduction, to a planned sequel which will explore the approaches to and the decisive contributions of contemporary earth science.

Chapter 5

THE MID-NINETEENTH CENTURY (1830-1870)

5.1 Economic, political and technological setting

The technological change of the Industrial Revolution
which had first begun in Britain was virtually completed
in that country by 1830. Furthermore, similar change was
spreading rapidly to other European countries. New
industrial towns and cities were beginning to grow and
urban life with its good and bad became the lot of an
ever-growing number of people. The forms of government
in Europe were, however, hardly suited to the new age of
industrialism. Policies derived from an earlier age more
often than not prevented the new industries from realizing
their full potentialities. Industrialism had conferred
wealth and power upon the growing commercial and industrial
interests (THOMSON, 1964, p. 331). These rising classes
stood to gain most from changes in forms of government,
and played an active role in politics in nineteenth-century
Europe. The political situation at this time was restless,
and in some countries prone to revolution (POUTHAS,
1964, p. 389). Political movements directed
towards change began to assume prominence after 1830.

The following decade saw revolutions in France, Belgium,
Germany, Italy, Switzerland, and Poland, and civil war
in Portugal and Spain. In Britain there was a strong demand
for reform. Political changes began to take place
creating conditions for an acceleration of industrial
expansion. In Britain the changes took the form of a
series of parliamentary measures among which were the
passing of the Reform Acts of 1832, the abolition of the
Navigation and Corn Laws, the adoption of free-trade
budgets under Peel and Gladstone and the Representation
of the People Act of 1867. These measures reflected the
transfer of power from the landed aristocracy to the rising
class of industrialists. At the same time that these
political changes were taking place, a new technical
development was occurring. Steam power, still the great
mover of industry, was being extended from production to
transportation. This development was to bring in a new
era, that of railways and steamships. These new political
and technical factors set into motion a rapid expansion
of industrial production and trade.

Influence on ideas

In such a setting the vision of progress through the
application of science and technology continued to motivate
many of the advocates of political change as well as a

section of intellectuals and scientists throughout Europe.
This spirit was especially characteristic of the period
between the French Revolution and the revolutions of
1848, subsequent to which somewhat different ideas came
to the fore (BURY, 1960, p. 15). Ideals of scientific
progressivism spread abroad from Europe: to colonies
such as Canada and to former colonies such as the
United States of America.

Organization of science in the mid-nineteenth century

Science, no less than other phases of life, was profoundly
affected by the new industrial age. The nature of science
as well as the outlook of scientists and the relationship
of science to the state and to national life underwent
a thoroughgoing change.

The British Association for the Advancement of Science

The contrast between the possibilities for improvement
through the applications of science and technology and
the actual low level of application of science to society
in the first decades of the century caused widespread
dissatisfaction with existing forms and organizations
of science. For one thing the Royal Society proved to
be, at least at first, ill-adapted to the new conditions.

This resulted in a strong movement of scientists directed towards forming an association outside the Royal Society. This movement involved leading scientists, among the most active of whom were Charles Babbage and David Brewster. Many of these scientists were members of the Royal Society. However, they felt that the latter had become a largely honorific and advisory body, ill-adapted to the principal needs of the time. The idea of the Association was inspired and to some degree modelled on the Deutscher Naturforscher Versammlung, in which von Humboldt had been active. The British Association, however, soon assumed a different character because it, unlike its predecessor, was surrounded by an expanding industrial economy. The supporters of the Association envisioned as a vital need of the nation some means of "strengthening the relationship between science and the public interests" and lending "guidance to the individual efforts of its members" (HOWARTH, 1931, p. 7 and 19). By a collective survey of nature the Association hoped to overcome to some degree the tendency in a period of rapidly growing science, towards specialization and the consequent inability of individuals to see the whole structure of science clearly. The Association attempted to create a proper organization to "direct...investigations to the points where an extensive survey thus generalized would indicate as the most important" (HOWARTH, 1931, p. 22).

These were the ideas, as given at its first meeting, which led
to the foundation of the British Association for the Advancement of Science in 1831. The Association had a profound
effect on science in Britain and its possessions in a number
of ways. This effect was felt through its grants for
scientific work, as well as through its committees and their
influence upon the Royal Society and upon the government,
and through its annual meetings and special lectures. In
the first decades of its existence a number of leading
scientists who had contributed to the scientific study of
the earth served as presidents of the Association. These
included Adam Sedgwick, Sir David Brewster, G. B. Airy,
William Whewell, Sir John Herschel, Sir R. I. Murchison,
Sir Charles Lyell, Sir George Stokes, Thomas Huxley, and
Lord Kelvin.

Technological trends

As for the technological trends of the period it has been
noted that "the mid-nineteenth century was not a period of
radical technical transformation that can be compared
with the eighteenth. It was rather one of steadily
improving manufacturing methods operating on an even
larger scale (BERNAL, 1965, p. 386). Furthermore, there
was "no violent urge for new devices in production.
There was on the other hand an ever-increasing need to

speed up communication and transport. The telegraph was
the first practical and large-scale application of the
new science of electricity..."Where the eighteenth century
had found the key to production, the nineteenth was to
find that of communication" (BERNAL, 1965, p. 386, 388).

5.2   The earth sciences in the mid-nineteenth century

The mid-nineteenth century brought many of the earth
sciences into their present-day forms. As regards scientific
content we have seen that some branches of the geophysical
sciences showed many of their modern features as early as
the Scientific Revolution, as did branches of the geological
sciences during the Industrial Revolution. All of these
were further modified in the mid-nineteenth century and
acquired organizational forms which have been retained in
many cases until the present day. Geomagnetism (which had
already developed into a recognizably modern form) went
through such a development in this period.

Geomagnetism since the Scientific Revolution

In Figure 3.2 we have seen how scientific activity in
geomagnetism varied through the Scientific and Industrial
Revolutions. As may be seen in the Figure, following the
peak of activity in geomagnetism which occurred during

the last phase of the Scientific Revolution (in the period
1650-1700), there was some slackening of the tempo in
geomagnetism. This is another example of "the early
eighteenth century pause" (BERNAL, 1965, p. 358) which
affected many branches of science, as well as geographical
exploration. With the beginning of the Industrial Revolution
and the economic and political changes which it brought
about, first in Europe and then elsewhere in the world, came
a new movement in geomagnetism. It shared the general
growth which began to occur in science. Interest in
navigation, exploration and geodetic mapping was high.
Great inland areas such as Russia and Africa were beginning
to attract attention. Geomagnetism, because of the part
played in it by the compass, was a part of this general
area of interest. By the mid-nineteenth century a new
interest in geomagnetism was stimulated by interest in
earth currents, because of their effects upon telegraphs.
The latter were a vital part of the opening up of new
land areas and were, therefore, influenced by the upsurge
of inland mapping in this period.

Maritime voyages of discovery

The sudden expansion of geomagnetism in the latter half
of the eighteenth century followed the pattern of earlier
centuries. The measurement and mapping of the magnetic

elements was closely tied to maritime navigation. However,
these were most often a part of newly-initiated voyages
of geographical discovery rather than a continuation of
traffic on the established trade routes of former times.
These voyages were largely government-sponsored expeditions,
with France and Britain the principal sponsors. New
technical developments resulted from this phase of exploration in which total intensity of the geomagnetic field
was measured rather than declination as formerly. In
this same half century the first map of inclination was
made, by J. C. Wilcke (HELLMANN, 1895, p. 13). The first of
the great scientific land expeditions was made in this period
by Alexander von Humboldt (during the period from 1798-1804)
to South America. As a result he constructed the first extensive map of geomagnetic intensity. As mentioned before this
expedition was first inspired by the scientific climate
in France at the time (Sec. 4.3). Sea voyages continued
as the principal type of expedition until about 1840. After
this time land expeditions gained in prominence, although
great effort was also expended on polar sea voyages.

The new character of land-based exploration

The development of geomagnetism in Russia is an example
of what was taking place everywhere within the spheres
of influence of the European powers. Markets and raw

materials became vital concerns of the industrial centres.
By mid-century, industrialism had reached centres in much
of Europe including Russia, where "not only had capitalist
industry taken root but the economic revolution had extended to foreign trade...railways were the chief material
achievements, their total length increasing from 1626 km.
in 1860 to 10,731 km in 1870" (VYVYAN, 1960, p. 380-1).
The pressure for markets and raw materials and the increase
of means of expanding into new territories, especially
the railway and telegraph, led to a vast effort of exploration
and military conquest which finally integrated Russia's
loosely held dependencies in Asia into an empire. This
drive to occupy territory fully (as contrasted with earlier
periods when coastal trading stations and fortresses
loosely connected by trading routes with the hinterland were
often the only presence of the European powers in dependencies) was characteristic of the decades following the
Industrial Revolution and was a prime motivation in all
the industrialized nations. This drive began with the
scientific voyages early in the century, which were most
often primarily for the purpose of geographical discovery;
later they came to centre on land expeditions such as those
in Africa and Asia. Often these enterprises were accompanied
by bitter international rivalry. In the Asian territories
later to be consolidated in the Russian Empire such rivalry

was intense, where "in competition with the British recon-
noitering from the Punjab, Russian explorers, savants and
diplomatic agents at least held their own" (VIVYAN, 1960, p.
386).

With the support of the Russian government Alexander
von Humboldt made an expedition to the Urals and to Siberia
in 1829-30. On these journeys he made geomagnetic measurements which were a valuable contribution to geomagnetism.
At a special session of the Academy of Sciences at St.
Petersburg in November 1829, Humboldt laid before
the Academy a proposal that a chain of meteorological
and magnetic observatories be established in the European
and Asiatic possessions of Russia. Humboldt's international
reputation gave weight to this suggestion and a committee
of the Academy was appointed to carry out the organization
of such a network. A similar proposal had been made by
Leibniz to Peter the Great in 1712 (KELLNER, 1963, p.
141) but was never carried out, although it would have
been technically feasible. The time was not right for
such a project to be seen as one of importance to the
nation. A similar enterprise was more favourably received
in the later period when exploration and occupation of the
eastern territories became an urgent national concern.
By 1835 a chain of stations was in operation from St.
Petersburg to Peking. Observations were also made at

Sitka, in Russia's territory in North America, from
1832-5. Although maritime expeditions and surveys of
course continued, the new development of this period was
the growth of land expeditions, magnetic surveys of the
territories of nations and fixed observatories. An earlier
Figure (1.3) shows the numbers of observatories in operation
from the seventeenth until the twentieth century. The
first phase of rapid growth marks the beginning of a
new era in earth science in which maritime interests were
to give way to land-based interests. Interest in land
areas was to dominate earth science until more than a
century had elapsed, shaping and developing the earth
sciences accordingly. A return to ocean-based science
in the present century has led to a new upsurge and new
directions in earth science, completing a cycle spanning
the period of modern science.

Stimulus and support for geomagnetism

We have seen how this subject has, in its rate of growth
and distribution as viewed through time, followed the
trends of general history. It is part of our object to
understand the exact ways in which the earth sciences
have been influenced by their scientific and historical
background and the ways in which they themselves have
influenced other branches of science and perhaps history

as well. It is instructive in this regard to look at two examples. The first is the British Association for the Advancement of Science, and the second is the Göttigen Magnetic Union.

The British Association and geomagnetism

We have already examined the character of this organization (Sec. 5.1), and seen that it was primarily an organization of scientists who were concerned about the fact that the traditional organization of science in Britain was failing to produce the growth and application of science which was clearly needed if the new powers brought by the industrial age were to be used effectively in the interests of the nation. The Association was a powerful one, with great influence; it did not limit itself to being a meeting place for scientists interested only in the dissemination of scientific ideas but also organized scientists for work on scientific projects which it felt were of paramount importance to the nation, providing research grants and other facilities for carrying out these projects.

Geomagnetism was named by the British Association in 1839 as "its paramount object" stating further in a memorandum to the British government in which official support for geomagnetic projects was being sought that "the time is...now arrived when all that is rude and inexact in the subject of terrestrial magnetism must give place to rigorous numerical statement and refined discussion...The object it is true,is to perfect a theory, but it is a theory, pregnant, as we shall see, with practical applications of the utmost importance....The laws of magnetism, under any circumstances important to a great maritime power, are every day acquiring additional interest by reason of the introduction of iron vessels into navigation..."

The Association established a committee for terrestrial magnetism in 1832, only one year after the formation of the Association. A survey of magnetic declination, dip and intensity of field over the territory of the British Isles was organized by a committee of the Association in 1834-8. Following urging by the British Association and the Royal Society, magnetic surveys at sea were conducted on the Sir James Clark Ross naval expedition to the Antarctic seas 1839-43. By the same order of Her Majesty's Government (1839), fixed observatories were established at St. Helena, Toronto, the Cape of Good Hope and Tasmania under the supervision of General Edward Sabine. The East India Company at the same time agreed to establish observatories at Madras, Bombay, and in the Himalaya mountains.

In 1845 the British Association held a geomagnetic conference at Cambridge which asked the government to continue the existing observatories established under the order of 1839 and to establish new observatories at Greenwich and Dublin.

Subsequent representations by the British Association and the Royal Society led to a succession of magnetic surveys on sea and land. Some of these are: voyages to Arctic waters, 1818-55; a second voyage to the southern oceans by the naval officer T. E. L. Moore and the army engineer Lieutenant Henry Clerk in 1845; a survey of the Indian Ocean co-ordinated with the magnetic observatory at Singapore (established in 1841 and run for four years) carried out by the army engineer Captain Elliot. The brothers Schlagintweit (of Prussia, who had done researches on earth physics in the Alps) surveyed British India in 1856-8. A re-survey of the British Isles was conducted in 1857, twenty years after the first survey.

A committee of the British Association, of which Balfour Stewart was the most active member, collected an important series of reports giving methods of interpretation of geomagnetic data, 1886-88.

The continued interest of the British Association in terrestrial magnetism shows that the subject was viewed by members of the scientific community as one of great scientific and national importance. These were large-scale projects, requiring government assistance for their execution. Thus some of the first large co-operative projects in science arose in the field of earth science, which thus was in the forefront in pioneering large-scale organized science. The history of the representations made by the British Association to the government indicates that it was always necessary for the scientists themselves to take the initiative in organizing and carrying out the projects. Furthermore, it is clear that scientists of a broad range of specialties took an interest in the work and carried out research in the subject. Often the government supported geomagnetism by assigning officers to geomagnetic work with military facilities made available for these projects. In the course of these developments, fostered by an association of scientists on one hand and the government on the other, a profession of geophysics began to emerge.

The Göttigen Magnetic Union

Carl Friedrich Gauss was a life-long friend of Alexander von Humboldt who had, since his return from his American journey in 1804, urged Gauss to take up an interest in terrestrial magnetism. Gauss was already involved in scientific studies of the earth, in his work on geodesy and figure of the earth. He had entered this field at a time when mapmaking was one of the great practical concerns of the inland states later to make up Germany. Gauss began to take part in geodetic surveys as early as 1802, and 1821-8 conducted a triangulation of the Kingdom of Hanover. Gauss was also interested in telegraphs, and he and Wilhelm Weber invented an instrument at Göttigen in 1833. Problems related to telegraph networks, influenced as they were especially in the early days by telluric current surges, elevated geomagnetism to a position of practical importance for telegraphy because of the relationship between geomagnetic and telluric effects. The discovery of the basic phenomena of electromagnetism, between 1820 (the date of Oersted's discovery of the magnetic effects of electric currents) brought magnetism into the forefront of scientific research. In this setting Gauss in 1831 took up the study of magnetism and developed a special interest in it. He obtained the support of Göttigen University in the establishment of an observatory, which was in operation by 1833. The observatory recorded the magnetic elements regularly, and instituted what were called "term days" on which up to 44 hours of continuous observations took place. Other centres were encouraged to make simultaneous observations, and by 1834, observatories were set up at Berlin, Frankfurt, Bavaria, Leipzig, Brunswick and Copenhagen. Göttigen became for a time a centre for magnetic research, with observatories all

over the world duplicating the instrumentation at that
centre and taking readings to conform with the term days
used at Göttigen. These observatories formed themselves
into what was called a "Magnetic Union". Gauss and
Weber began in 1837 to edit a periodical at Gottigen
to publish the results of these co-ordinated efforts.
The periodical was published for 1836-41 in six volumes.
In 1837 Gauss and Weber invented the bifilar magnetometer
and in 1838 Gauss published the first comprehensive
quantitative analysis of the earth's magnetic field.
The development of Gauss' method of analysis was one of
the great achievements in geomagnetism. We have seen the
influences arising from Gauss' time and place which led him
as a leading mathematician into the field of terrestrial
magnetism. This was brought about to a considerable
extent by Alexander von Humboldt who acted personally
to interest Gauss in the subject. In addition Humboldt
used his great international influence (as a scientist
and a diplomat) to facilitate the formation of the
Magnetic Union in order to further the resultant co-operation
and collection of data. Thus we see a case where a
naturalist influenced basic physics, setting the background
for one direction of Gauss' mathematical work. Is is
interesting to note that Gauss' involvement in geodesy
also influenced other facets of his mathematical work,

for example in his development of the theory of curved
surfaces (DUNNINGTON, 1955, p. 163-6).

The growth of geomagnetic observatories

This is a matter which is particularly suitable as an example
of the growth of an earth science and the relationship
of this growth to general history. Terrestrial magnetism,
referred to as "this important subject" in the Presidential
Address to the British Association for the Advancement
of Science in 1846, was for a long time held in high
regard by scientists and non-scientific administrators
alike. Magnetic observatories, affording as they did a
continuity of observation, were felt to be the mainstay of
the subject. Of course magnetic maps over sea and land
areas, as well as individual observations have been very
important components of geomagnetism from its earliest
days. However, the continuous or frequently repeated
recordings from a network of fixed sites were long regarded
as the central body of data for geomagnetism.

The first geomagnetic observatory was founded in
Paris in 1667. Figure 1.3 shows the number of observatories
operating during any particular half-century from then to
1800 and during any particular decade from 1800 to the
present time. The data were compiled from FLEMING and
SCOTT (1943, pp. 97-108, 171-182, 237-242; 1944, pp. 47-

52, 109-118, 199-205, 267-269; 1948, pp. 199-204; FANSELAU,
1965). The Figure indicates several phases of growth.
The first begins in the closing years of the Industrial
Revolution, reaching its maximum rate of growth in the
mid-nineteenth century period. The number of observatories
grew during this phase to a total of about forty, and
oscillated about this value for several decades. This
upper limit of the growth curve represents the cumulative
result of the upsurge of overseas trade and mapping of
inland territories which developed as the Industrial
Revolution matured and graded into the subsequent period.

The foregoing sections describe how scientific
and national motivations led individual scientists, influential scientific societies, governments, and trading
companies to support a worldwide network of geomagnetic
observatories during this period of initial and rapid
growth. This support levelled off by about 1850, but the
observatory network had been sufficiently well established
to maintain the level that had been reached. A new
period in the history of science, the Age of Imperialism
(Sec. 1.3), began to show its influence on geomagnetism
during the decade 1870-80 and initiated a new phase of
growth. The momentum of the previous period had died off,
but the new conditions brought a change to geomagnetism.
The First Polar Year, part of the international geophysical

tradition which consolidated during the Age of Imperialism
(Sec. 1.6), occasioned the sudden rise evident on Figure
1.3. The Year (described by HEATHCOTE and ARMITAGE, 1959)
was part of an internationally planned scientific effort
concentrated on the north polar regions. Observations of
terrestrial magnetism and electricity, aurorae, as well
as weather and climate were carried out by expeditions
from a dozen countries. The momentum of this period was
such that the number of observatories did not drop back
after the Year, but continued to grow throughout the
period. This is an interesting case in the history of
science since, in contrast to general history where a
distinct break occurs, it is not evident in all sciences
that a distinction can be made between the first and the
second half of the nineteenth century. This distinction
occurs often enough that BERNAL (1965) demarcates such
a period ("the late nineteenth century") although remarking
(p. 400) that "it is a difficult period to demarcate,
particularly in science...for the change here was a gradual
one without any marked break in continuity". It is not
surprising that the break can be seen in Figure 1.3 in earth
science, tied as the earth sciences are to development and
exploration and therefore to the imperialist expansion of
the late nineteenth century. It is interesting to note
that until 1945, except for a special international effort,

the Second Polar Year (for which see Sec. 1.3, describing
the provisional nature of that project) no substantial
change occurred in the number of observatories. Following
World War II, a new period began, that of government
science (Sec. 1.3). A growth phase was initiated, culminating in the International Geophysical Year.

To summarize, we may note that growth in the number
of geomagnetic observatories occurred in three rapid
spurts, centred on 1830, 1885 and 1955. The rate of
growth was most rapid in the last, almost 10 observatories
per year at its maximum. Next largest is in the first part
of the nineteenth century, with a maximum rate of two or
three per year. Growth was slower in the second half of
the century perhaps one per year (if the first polar year
in which stations operated only for the duration of that
project are excluded).

The study of the earth's shape and gravity in the nineteenth century

The history of this subject is a good illustration of the
fact that social developments in this period continued
to be closely intertwined with developments in earth
sciences, as they had been in earlier periods. This new
age of industrial production presented new problems to
science. The capability of the latter had increased

during and since the Industrial Revolution, making possible advances of a new degree beyond those of the previous periods.

The spirit of the earlier age of discovery (the Scientific Revolution) had fostered the rebirth of the feeling that the earth and its place in the universe should be considered in a fully scientific manner, as the ancients had once done. Some of the principal problems set by the age of maritime expansion were related to the motion, shape, and gravity of the earth. In this atmosphere some of the greatest scientists of the time took up these problems as their own. A chain of mathematical syntheses, beginning with those of Isaac Newton in the seventeenth century and ending with the works of Pierre Laplace at the end of the eighteenth century carried the science of the earth through an important phase. The body of science thus created at the end of the eighteenth century had developed as far as was possible, given the general level of science at the time. The problems set by a new age (the mid-nineteenth century) and supported by the more refined theories of the mechanics of materials, which were developed in this period in the course of solution of engineering problems (which had arisen from the Industrial Revolution) were ones which led the subject onto new pathways. Previously the subject had been dominated by ocean navigation and the problems arising therefrom. Following the Industrial Revolution, new problems related to exploration, transport, and mapping on land began to assume importance. These developments were related to the upsurge in production of industrial raw materials and the expansion of the transport system in the form of canals and railways. In Europe this was also the period of consolidation into modern states and the beginning of large scale movements of people into other continents, in new territories such as America, Canada and Siberia. All of these developments required accurate geodetic surveying and mapping on land. That the general scientific community viewed these concerns as important is shown by the resolutions passed by the British Association for the Advancement of Science, which pressed the government many times from 1838 on to extend the ordinance survey maps at home and abroad and to keep mining records and surveys on file (HOWARTH, 1931, p. 212, 213).

Mechanics of materials in the earth's interior

The mathematical theories of the shape of the earth up to the end of the eighteenth century treated only the very simple problem, in which the material composing the body of the earth was considered to have flowed as a perfect fluid into an equilibrium position under the influence of rotation and gravitation. Agreement was found

between these theoretical results and experimental measurements such as geodetic arcs and pendulum observations of
gravity. This agreement was interpreted as meaning that
the earth had been fluid, perhaps in its early stages.
In problems such as the analysis of ocean tides, the
treatment of the solid earth had not progressed beyond the
simplest case, treating it as a rigid and inert region
playing no part in the tidal process. A contrasting view
of the earth's interior was widely held at the same time.
In the geological sciences at the beginning of the nineteenth
century "the scientific authorities...regarded it as an
accepted fact that the earth's nucleus was molten, and was
surrounded by a comparatively thin crust" (VON ZITTEL, 1901,
p. 177).

Mathematical and physical studies of the earth
gradually began to establish a more balanced view. A memoir
of William Thomson (later to be Lord Kelvin) in 1863 showed
that the hypothesis of complete internal fluidity is
untenable, as is the opposite extreme, the hypothesis of
rigidity. He departed from the traditional approach of
treating the earth as rigid in tidal problems, assuming
an elastic earth, and showed that a body tide was to be
expected as well as an oceanic tide and that because the
earth itself is a yielding body, the oceanic tide should
be less than that indicated by theory assuming a rigid

earth. By comparison with tidal data he was ablt to conclude that the "tidal effective rigidity" of the earth was at least that of glass. This analysis was not the first but came almost a quarter of a century after the first physical calculation to upset the simple ideas of all fluidity or all solidity which had been the two prevalent models previously. This first physical attack on the problem was made by the Cambridge physicist, W. Hopkins (1839), "to whom is due the grand idea of thus learning the physical condition of the interior from phenomena of rotatory motion presented by the surface" (THOMSON, 1863). Hopkins investigated the amount of luni-solar precession and nutation, assuming the earth to consist of a solid spheroidal shell filled with fluid. He found that the precession and nutation for a liquid-solid model differ from that for a completely solid or a completely liquid sphere by an amount dependent on the thickness of the crust or solid shell enclosing the fluid. By comparison with the observed motions of the earth he concluded in later memoirs (HOPKINS, 1840, 1842) that the thickness of the solid crust of the earth lies in the range "800 to 1000 miles". This estimate gives a picture which is not too far different from our present-day view of the earth's interior, although we now know that the solid layer extends to a depth of about 1800 miles.

This period in history is marked by the appearance of mathematical and physical analyses of the reaction of the earth's crust and interior to stresses of various intensities and time scales. The earliest work, by Lamé (1854) and then H. Resal (1855), investigated the stresses in a rotating elastic shell, and applied it to the earth. These researches postulated a stress system existing in the earth's crust, "thus initiating those investigations in terrestrial physics which have been still further advanced by Sir William Thomson (Lord Kelvin), G. Darwin, Chree and others" (TODHUNTER, 1893, p. 389).

Anelastic behaviour of the earth

The great interest in metals during the technological age which developed in the nineteenth century led to the discovery of anelastic effects. Viscosity and after-strain in metals was discovered by A. T. Kupffer at the Physical Observatory St. Petersburg where "Kupffer's discovery of viscosity and after-strain in metals dates at least from 1852" (TODHUNTER, 1893, p. 513). The quantitative theory of plasticity began with Tresca and then St. Venant in 1868 "following a long course of experiments on the punching and sequeezing of metals (DARWIN, 1879, p. 29). The dependence of mechanical properties of matter on the time-scale of applied forces received more precise

formulation in the physical ideas of Clerk Maxwell in
1866 which in turn received mathematical formulation by
J. G. Butcher in 1876, and by William Thomson in 1878.
These models found their first application to the earth
in 1879 by G. H. Darwin.

Problems raised by geodesy and mapping

The scientific upsurge before 1700 with its strong interest
in global navigation and the mapping of the earth had
ensured that geodetic surveys of high accuracy were
carried out, often with the most advanced scientific
support available at the time. Interest continued into
the mid-nineteenth century and geodetic projects were
supported by academies of science, national observatories
and other leading bodies controlling the development
of science. Thus the best scientists of the time gave
their consideration to the results. Such detailed
scrutiny resulted in a number of advances in understanding
the earth. For example, it had been recognized well before
the mid-nineteenth century period that the attraction of
mountains on the plumb line could cause errors in geodetic
surveys. Calculations of the magnitude of such effects
had been made by Pierre Bouguer in 1749, D'Alembert in
1756, and Henry Cavendish in 1768 (TODHUNTER, 1873, pp.
248, 284, 448). These problems continued to be of

interest in the nineteenth century because of the social
importance of land mapping at this time.

A growing interest in the depths of the earth,
in an age in which geology had become established,
inspired interest in the effects observed by Pierre
Bouguer and de la Condamine at Chimborazo in 1737 and
1738 (Sec. 2.6). Bouguer had been somewhat dissatisfied
with his results, suggesting that the Cordillera area is
unsuitable for such an experiment. For example, internal
cavities in an extinct volcano like Chimborazo might decrease
the mean near-surface density (MACKENZIE, 1865, p. 43).
He suggested that a more compact hill in England or France
might provide a more satisfactory experiment. Such a
topographic feature (the mountain Schehallien) was used by
Henry Cavendish and Charles Hutton in 1778 to provide the
first accurate calculation of the earth's mean density from
the gravitational effects of a surface feature.

A comparable experiment even more free of the above
mentioned objections is to measure the vertical gradient
of gravity in a mineshaft. G. B. Airy attempted such an
experiment in 1826 in the Dolcoath mine in Cornwall, in
collaboration with William Whewell (AIRY, 1896, p. 66-68).
This experiment was inconclusive but greater success was
met with in 1854, when he measured a vertical gradient of
gravity in the Harton Colliery and calculated the mean

density of the earth with greater precision than that
obtained in the earlier experiment (conducted 76 years
previously) of Cavendish and Hutton. Such experiments have
been repeated, with increasing accuracy, at intervals down
to the present time (Figure 5.1).

The attention given by Airy to this problem
illustrates another feature of the time. Leading scientists
whose main work and reputation lay in other fields often
took up work on earth science. This work appeared to them
as exploration of a vital area of nature. G. B. Airy was one
such scientist: he was productive with 518 printed papers
and eleven books. He was also distinguished in many fields
and was the Astronomer Royal for forty-six years (from 1835
to 1881). Airy is an example of a talented scientist with
broad interests and a strong attraction to a whole range of
scientific and technical subjects (as indicated by the list
of titles of publications in AIRY, 1896, pp. 373-403). Of
these, sixty per cent are on astronomy, twenty per cent on
general physics, mathematics and engineering, and twenty per
cent on earth science. The latter twenty per cent represents
almost 100 papers devoted to the science of the earth.

Further discoveries regarding the earth's interior

The deflection of the vertical became of widespread concern in
this age of new empire, when mapping over large areas of the

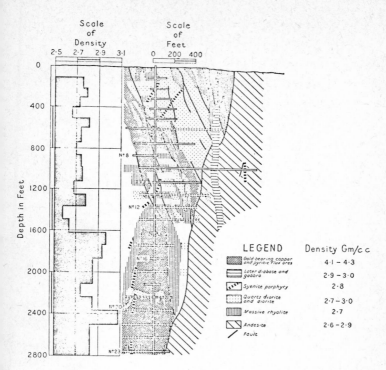

Fig. 5.1   Variation of gravitational acceleration and
             derived densities in a mine.   From these
             the mean density of the earth can be calcu-
             lated.   From MILLER and INNES, 1953.   Appli-
             cation of gravimeter observations to the
             determination of the mean density of the earth
             and of rock densities in mines. Publications
             of the Dominion Observatory, Ottawa, xvi(4),
             by permission of the authors.

earth became a vital activity of many nations. The problem was expressed thus at the time: "if in mapping a country we calculate the latitudes with a wrongly-assumed curvature, or from the data given by plumb-lines which are influenced by disturbing causes, known or unknown, the results must be affected" (PRATT, 1860, p. 105). These remarks followed the first attempts at precise surveys at the base of the world's greatest mountain system, the Himalayas, during the first geodetic survey of the great Arc of the Meridian in India. They were made by J. H. Pratt, then Archdeacon of Calcutta, who did much calculation during the course of these geodetic operations. He went on to say that "the irregular character of the surface of the Earth over large tracts of country, consisting of mountain and valley and ocean, may in some instances have a sensible effect, by presenting an excess of deficiency of attracting matter, upon the position of the plumb line, in such a way as to derange delicate survey operations. Hindostan affords a remarkable example of this as the most extensive and the highest mountain-ground in the world lies to the north of that continent and an unbroken expanse of ocean stretches south down to the south pole. Both these causes by opposite effects make the plumb line hang somewhat northerly of the true vertical (PRATT, 1860, p. 49). In 1855 Pratt calculated the amount of this affect at the most northerly station of the Great Indian Arc to be 34" on the basis of gravitational theory (PRATT, 1855; 1860, p. 56).

An unexpected discrepancy appeared in which the observed
value was some 7 " less than expected from Pratt's calculation. This small deviation by itself would likely be too
small to be given much weight; however, similar discrepancies
with the same sense were observed at other stations (PRATT,
1860, p. 60). Pierre Bouguer had observed and described
similar effects in the Andes a century earlier. By this time
the effects of topographic features on the plumb line were
better understood. Pratt in 1855 had better topographic information available than Bouguer had in 1737 and he was much less
uncertain than Bouguer had been of the effects to be expected
from the terrain. Thus Pratt's calculation made it necessary
to consider conditions <u>below</u> the mountains as a possible explanation of the discrepancies between his calculations and
the measurements. G. B. AIRY (1855) suggested that a solid,
thin crust overlying a fluid substratum and in hydrostatic
equilibrium in it would, if the crust were thickened below
mountains, by virtue of the density deficiency as compared with
strata lying to the side of it thereby created, explain reasonably well the amount of deficiency in deflection of the vertical.
In the same year PRATT (1855) made the alternative suggestion
that even if the solid crust were thick, "the mountains may
have drawn their mass from the regions below through a considerable depth, by an extension and small expansion of the
matter in those lower regions...trifling deviation in the

density...(from that required for fluid-equilibrium) if it prevails through extensive tracts may have a sensible effect upon the plumb line" (PRATT, 1860, pp. 56-71).

These two ideas appeared at first to be in opposition, each representing one side of a controversy. The first, that of Airy, was based on the idea of a thin crust underlain by fluid, an idea which was generally popular with geologists. The second, that of Pratt, was based on the idea of a thick upper solid zone in the earth's interior, and idea that was more generally held by physicists such as William Thomson than by geologists. PRATT (1860, p. 88) argued strongly for this view saying that "the result of the whole proves that the crust must be very thick: and, as Mr. Hopkins' calculation appears to be free from objection, and in fact to be the only one on which any reliance can be placed, we may conclude that the thickness is at least 1000 miles......The present form of the surface in mountains, table-lands, continents, and oceans has been, no doubt, acquired from a process of expansion and contraction which the crust has undergone during the ages since it was first con-solidated....We may, therefore, fairly conclude...that the present varieties of the Earth's contour have grown from this cause, and have not arisen in any way from the operation of hydrostatic principles." Pratt's view of

mountain formation as due to uplift driven by heat is
reminiscent of James Hutton. Due to the fact that precise
definitions of solid and fluid had not yet appeared,
and that the possibility of the same material acting as
either one depending on the time scale of application of
the deforming forces had not yet received precise formulation, it was not recognized that these ideas were not
in fact in contradiction. Today we know that in various
parts of the earth one or the other or a combination of these
two extreme models applies.

5.3 Conclusions

The earth sciences continued into the mid-nineteenth
century with their general style of development closely
related to that of the technological and social developments
of the time. The continuation of this relationship will
be explored in a sequel.

APPENDIX 1

ANALYSIS OF DATA ON THE GRAVITATION AND FIGURE OF THE EARTH

In an earlier section (3.3) it was pointed out that
Figure 3.3 is constructed from two sets of data, somewhat
different, but which overlap in the period 1800-1825. The
first set of data gives N (total number of pages (periodical
and in book form) for a given time period, while the second
gives P (the total number of published papers in a given
time period. In the overlap period, N is found to be
proportional to P, i. e. N = kP where k is a constant.
The value of k is in fact about 10. If we make the assumption,
resaonable in the absence of further knowledge about it,
that k remains about the same until 1900, we can by a
simple multiplication reduce the two sets of data to a
common basis. Thus an index was developed for the rate of
growth of literature on the earth's gravitation and figure
which is applicable over the whole time period treated
in Figure 3.3.

APPENDIX 2

LIST OF AUTHORS REFERRED TO IN FIGURE 3.9

AGRICOLA (Georg Bauer) (1494-1555) - Saxon physician and
professor of chemistry at Chemnitz; His "De re metallica"
was the basis for many of the later developments in
metallurgy and mineralogy.

ANAXIMANDER OF MILETUS (c. 610-547 B.C.) - one of the
earliest philosophers of the Ionian school, and the fragments
of his writings which remain indicate that he made advances
towards serious paleontological investigations.

ARISTOTLE (384-322 B.C.) - founder of the Lycium in Athens
(335 B.C.); wrote on metals, minerals, rocks, earthquakes,
springs and rivers.

AVICENNA (980-1037) - Arabian physician; wrote on the cause
of mountains, and on fossils.

BERZELIUS, Jakob (1779-1848) - Swedish chemist; was the
first to present a classification of minerals based on
their chemical composition; introduces silicates into
mineralogy, and recognized isomorphism.

BRONGNIART, Alexandre (1770-1847) - French mineralogist
and zoologist; was associated with Cuvier in the studies
which established new standards and new methods in
stratigraphic geology.

BRUNO, Giordano (1548-1600) - born in Naples, led a
wandering life throughout Europe; was burned at the stake for
heresy in Rome in 1600; wrote on the changing relative
levels of land and sea and their global relationships, as
well as on volcanoes.

CHU-HSI - 12th century philosopher in China; Wrote about
the process of petrification.

CUVIER, Léopold Chrétien Frédéric Dagobert, Baron (1769-
1832) - the great comparative anatomist who brought a
rational basis to paleontology.

HAÜY, Abbé René Just, (1793-1822) - French mineralogist; was the founder of the science of crystallography and thus of systematic mineralogy.

HOOKE, Robert (1635-1703) - English naturalist, engineer and experimentalist; for his many contacts with our subject, see various sections, this book; wrote on the nature of fossils, and on earthquakes and the evolution of landmasses.

HUTTON, James (1726-1797) - See section 4.2

KAZWINI, Mohammed - Arabian writer of the thirteenth century; wrote on the nature of fossils, and on earthquakes and the elevations of land and sea.

KIRCHER, Athanasius - seventeenth century writer on earth-quakes and the origin of springs and rivers.

LHUYD, Edward (1660-1709) - author of a catalog of English fossils.

LISTER, Martin - seventeenth century English writer on fossils and stratigraphy.

LI TAO-YUAN - 6th century geographer in China; described fossil fish.

LO HAN (fl. 375) - official of the Chin dynasty; recognized and recorded fossils as the remains of living things.

LUCRETIUS (c. 99-55 B.C.) Roman writer who wrote on minerals, stones, volcanoes and earthquakes.

LYELL, Sir Charles (1797-1875) - probably accomplished more in the advancement of geological knowledge than any other one person; the world-wide influence of his treatises and textbooks definitely established the principle of uniformitarianism.

PALISSY, Bernard (1514-1589) - French artist and naturalist; suffered imprisonment and death in prison because of his beliefs; was among the early writers (1580) who recognized the true nature of fossils, and who understood that springs and rivers come from well known surface phenomena such as snows and rain. In his day, this was a revolutionary idea.

PERRAULT, Pierre - 17th century Franch scientist; wrote on the origin of springs and rivers.

PLINY THE ELDER (A.D. 23-79) - an officer in the imperial service of Rome; made an outstanding contribution to the knowledge of minerals, and in recording a great body of information from older writers, which would oterwise have been lost.

RAY, John (1627-1705) - celebrated English naturalist who wrote on the origin of fossils and mountains.

SHEN KUA - 11th century astronomer engineer and official in China; described many fossils.

SMITH, William (1769-1839) - English civil engineer; credited with the first adequate concept of the relations of strata to each other and of the significance of fossils in stratigraphic geology; Smith's ideas and maps, rather than his scanty writings, justify his pre-eminent place in the history of the science.

STENO, Nicholas (1638-1687) - born in Denmark, but carried out most of his work in Florence; wrote on the formation of mountains.

STEVIN, Simon (1548-1620) - engineer and government official in Belgium and Holland; wrote on mechanics and hydrostatics; also wrote on physical geology and the evolution of the earth's surface.

STRABO (c. 63 B.C.- c. 20 A.D.) - Greek geographer and historian; wrote on the elevation and subsidence of land as well as on volcanoes, earthquakes, ores and minerals, and alluvial deposits.

THEOPHRASTUS (c. 370-287 B.C.) - successor to Aristotle as head of the Lycium in Athens; wrote a treatise on minerals and rocks.

VALLISNIERI, Antonio (1661-1730) - president of the University of Padua; recognized the true nature of fossils; described mountain formations and springs.

VARENIUS, Bernhard (1622-1650) - German physician and geographer; laid the foundations for geography as a science other than the simple description of towns, nations and rivers; wrote on physical geology and on physical processes within the earth.

VINCI, Leonardo da (1452-1519) - Italian painter, sculptor, architect and naturalist; made a profound impression upon art, engineering and science; was one of the first in Renaissance Europe to recognize the true nature of fossils; recognized the role of aqueous forces in shaping mountainous terrain, and understood the origin of springs and rivers in such areas.

WERNER, Abraham Gottlob (1749-1817) - was for many years the most famous of the professors in the Freiberg School of Mines. His Neptunist school may have retarded geological thought, but his inspiring teaching and his earnest effort to classify all data did even more to advance it.

WOODWARD, John (1665-1728) - professor of medicine, contemporary of Hooke and Ray; wrote on fossils.

YEN CHEN-CHING - writer in China; mentioned fossils of bivalves (about 770 A.D.).

REFERENCES

ADAMS, F. D., 1938. The birth and development of the geological sciences. Williams and Wilkins, Baltimore, 506 pp. Paperback reprint, 1954, Dover, New York, 506 pp.

AIRY, G. B., 1855. On the computation of the effect of the attraction of mountain masses, as disturbing the apparent astronomical latitude of stations in geodetic surveys. Transactions of the Royal Society of London, B, 145: 101-103.

AIRY, G. B., 1896. Autobiography. University Press, Cambridge, 414 pp.

ARMITAGE, A., 1966. Edmund Halley. Thomas Nelson and Sons Limited, London, 220 pp.

BAILEY, E. B., 1950. James Hutton, founder of modern geology. Proceedings of the Royal Society of London, Edinburgh, 63B: 357-369.

BAILEY, E. B., 1967. James Hutton, the founder of modern geology. Elsevier, London, 161 pp.

BAKER, J. N. L., 1937. A history of geographical discovery and exploration. Second edition. Harrap, London, 553 pp.

BERNAL, J. D., 1965. Science in history. Watts, London, 1039 pp.

BERNAL, J. D., 1971. Science in history. MIT Press, Cambridge, Massachusetts, 1: 1-364; 2: 365-694; 3: 695-1008; 4: 1009-1330.

BIRCH, T., 1756-7. The history of the Royal Society of London, 2: 501, 4: 558.

BLAKE, R. M., DUCASSE, C. J. and MADDEN, E. H., 1960. Theories of scientific method: the Renaissance through the nineteenth century. University of Washington Press, Seattle, 346 pp.

BONDI, H. and GOLD, T., 1948. The steady-state theory of the expanding universe. Monthly Notices of the Royal Astronomical Society, 108: 252-270.

BONDI, H., 1961. Cosmology. The University Press, Cambridge, 182 pp.

BURY, J. P. T., 1960. Introductory summary. In: J.P.T. BURY (Editor), The new Cambridge modern history. The zenith of European power 1830-70. University Press, Cambridge, X: 1-21.

CAJORI, F., 1928. Newton's twenty year delay in announcing the law of gravitation. In: Sir Isaac Newton 1727-1927. Williams and Wilkins Company, Baltimore, pp. 127-188.

CHAPMAN, S. and BARTELS, J., 1940. Geomagnetism. Clarendon Press, Oxford, 1049 pp.

CLARK, J. G. D., 1969. World prehistory: a new outline (2nd edition). Cambridge University Press, London, 331 pp.

COLE, W. A. and DEANE, P., 1965. The growth of national incomes. In: H. J. HABAKKUK and M. POSTAN (Editors), The Cambridge economic history of Europe, VI, the industrial revolutions and after: incomes, population and technological change. Cambridge, pp. 1-59.

CREW, H. and DE SALVIO, A., 1914. (Translators) Dialogue concerning two new sciences, Galileo Galilei. Northwestern University Press, Evanston, 300 pp.

CROWTHER, J. G., 1942. The social relations of science. MacMillan, New York, 664 pp.

CROWTHER, J. G., 1948. Conservation and utilization of resources. Unpublished report to UNESCO.

CROWTHER, J. G., 1955. Science unfolds the future. Fredrick Muller Limited, London, 252 pp.

CROWTHER, J. G., 1960a. Francis Bacon. The Cresset Press, London, 362 pp.

CROWTHER, J. G., 1960b. Founders of British Science. The Cresset Press, London, 296 pp.

CROWTHER, J. G., 1962. Scientists of the industrial revolution. Cresset Press, London, 365 pp.

CROWTHER, J. G., 1967. Science in modern society. The
    Cresset Press, London, 403 pp.

DARWIN, G. H., 1879. On the bodily tides of viscous and
    semi-elastic spheroids, and on the ocean tides upon
    a yielding nucleau. Philosophical Transactions of
    the Royal Society of London. I. 170: 1-35. (also in
    DARWIN, G. H., 1908, 2: 1-35).

DAVIS, H. T., 1941. The theory of econometrics. The Principia
    Press, Bloomington, Indiana, 482 pp.

DEACON, M., 1971. Scientists and the sea. Academic Press,
    London, 445 pp.

DEANE, P. and COLE, W. A., 1962. British economic growth
    1688-1959. University Press, Cambridge, 348 pp.

DEDIJER, S., 1968. Early migration. In: W. ADAMS (Editor),
    The Brain Drain. MacMillan Company, New York, pp. 9-28.

DE MOIDREY, J., 1904. Note sur quelques anciennes
    declinaisons. Terrestrial Magnetism and Atmospheric
    Electricity, 9: 18-24.

DE SITTER, W., 1932. Relativity and modern theories of the
    universe. In: M. K. MUNITZ (Editor), 1957, Theories
    of the Universe. The Free Press of Glencoe, New York,
    pp. 302-319.

DUHEM, P., 1954. The aim and structure of physical theory
    (translation from second edition of La theorie
    physique: son objet, sa structure, 1914). Princeton
    University Press, Princeton, 344 pp.

DUNNINGTON, G. W., 1955. Carl Friedrich Gauss: titan of
    science. Hafner, New York, 479 pp.

EINSTEIN, A., 1952. The principle of relativity. Dover
    Publications, New York.

FANSELAU, G., 1965. List of geomagnetic observatories. IAGA
    Bulletin No. 20, IUGG Publication Office, Paris, 110 pp.

FARRINGTON, B., 1936. Science in antiquity. Oxford University
    Press, London, 256 pp.

FERSMAN, A., 1958. Geochemistry for everyone. Foreign
    Languages Publishing House, Moscow, 454 pp.

FLEMING, J. A. and SCOTT, W. E., 1943. List of geomagnetic observatories and thesaurus of values. Terrestrial Magnetism and Atmospheric Electricity, I-48: 97-109, II-48: 171-182, III-48: 237-242.

FLEMING, J. A. and SCOTT, W. E., 1944. List of geomagnetic observatories and thesaurus of values. Terrestrial Magnetism and Atmospheric Electricity, IV-49: 47-52, V-49: 109-118, VI-49: 199-205, VII-49: 267-269.

FLEMING, J. A. and SCOTT, W. E., 1948. List of geomagnetic observatories and thesaurus of values. Terrestrial Magnetism and Atmospheric Electricity, VIII-53: 199-204.

GEIKIE, A., 1897. The founders of geology. MacMillan, London, 297 pp.

GILBERT, W., 1600. De magnete (Mottelay translation, 1893). Reprinted in 1958 by Dover Publication, New York, 368 pp.

GILLMORE, C. S., 1971. Coulomb and the evolution of physics and engineering in eighteenth century France. Princeton University Press, Princeton, New Jersey, 328 pp.

GOFFMAN, W., 1971. A mathematical method for analyzing the growth of a scientific discipline. Journal of the Association for Computing Machinery, 18(2): 173-185.

GRANT, R., 1852. History of physical astronomy. Johnson Reprint Corporation (reprinted 1966), New York, 637 pp.

GRÜNBAUM, A., 1961. The genesis of the special theory of relativity. In: H. FEIGL and G. MAXWELL (Editors), Current Issues in the Philosophy of Science. Holt, Rinehart and Winston, New York, pp. 43-53.

GUYE, S. and MICHEL, H., 1971. Time and space measuring instruments from the 15th to the 19th century. Praeger Publishers, New York, 289 pp.

HAGNER, A. F., 1963. Philosophical aspects of the geological sciences. In: C. C. ALBRITTON (Editor), The Fabric of Geology, Addison-Wesley, Reading, Massachusetts, pp. 233-241.

HALLEY, E., 1683. A theory of the variation of the magnetical compass. Philosophical Transactions of the Royal Society of London, 13: 208-221.

HALLEY, E., 1686. A discourse concerning gravity. Philosophical Transactions of the Royal Society of London, 16: 3-21.

HALLEY, E., 1692. An account of the cause of the change of the variation of the magnetical needle. Philosophical Transactions of the Royal Society of London, 22: 563-578.

HARRADON, H. D., 1943. Some early contributions to the history of geomagnetism. Terrestrial Magnetism and Atmospheric Electricity. I-48: 3-17; II,III-48: 79-91; IV-48: 127-130; V-48: 197-200; VI-48: 200-202.

HARRADON, H. D., 1944. Some early contributions to the history of geomagnetism. Terrestrial Magnetism and Atmospheric Electricity. VI-49: 185-198.

HARRADON, H. D., 1945. Some early contributions to the history of geomagnetism. Terrestrial Magnetism and Atmospheric Electricity. VIII-50: 63-68.

HEATHCOTE, N. H. de V. and ARMITAGE, A., 1959. The first international polar year. Annals of the International Geophysical Year, 1: 6-100 (in English), and 105-205 (in French).

HELLMANN, G., 1895. Die altesten Karten der Isogonen Isoklinen Isodynamen. Kraus reprint, Nendeln/Liechtenstein, 1969, 23 pp.

HELLMANN, G., 1899. The beginnings of magnetic observations. Terrestrial Magnetism and Atmospheric Electricity, IV(2): 73-86.

HOPKINS, W., 1840. Researches in physical geology - second series. Philosophical Transactions of the Royal Society of London, pp. 193-208.

HOPKINS, W., 1842. Researches in physical geology: third series. Proceedings of the Royal Society, pp. 367-369.

HOWARTH, O J. R., 1931. The British Association for the advancement of science: a retrospect 1831-1931. British Association, London, 330 pp.

KELLNER, C., 1963. Alexander von Humboldt. Oxford, London, 246 pp.

KOSMODEMYANSKY, A., 1956. Konstantin Tsiolkovsky, his life and work. Foreign Languages Publishing House, Moscow, 100 pp.

LENZEN, V. F. and MULTHAUF, R. P., 1965. Development of gravity pendulums in the 19th century. Contributions from the Museum of History and Technology: paper 44, bulletin 240, pp. 304-347.

LILLEY, S., 1948. Men, machines and history. Cobbett Press, London, 239 pp.

LILLEY, S., 1966. Men, machines and history (2nd edition). International Publications, New York, 352 pp.

LLOYD, H. A., 1957. Mechanical timekeepers. In: C. SINGER et al.(Editors), A history of technology, Clarendon Press, Oxford, 3: 648-675.

LYELL, C., 1875. Principles of Geology (12th edition). John Murray, London, 1: 655 pp., 2: 652 pp. (1st edition published in 1830).

MACGREGOR, M., 1950. Life and times of James Hutton. Proceedings of the Royal Society of London, Edinburgh, 63B: 351-356.

MACKENZIE, A. S., 1865. The laws of gravitation. American Book Company, New York, 160 pp.

MASON, S. F., 1953. A history of the sciences. Routledge and Kegan Paul, London, 520 pp.

McGRAW-HILL, 1960. Encyclopedia of science and technology, 4: 345 pp.

McINTYRE, D. B., 1963. James Hutton and the philosophy of geology. In: C. C. ALBRITTON (Editor), The Fabric of Geology. Addison-Wesley, Reading, Mass., pp. 1-11.

McKELVEY, V. E., 1963. Geology as the study of complex natural experiments. In: C. C. ALBRITTON (Editor), The Fabric of Geology. Addison-Wesley, Reading, Mass., pp. 69-74.

MENDELEYEV, D., 1897. The principles of chemistry (second English edition). Longmans, Green and Company, London, 1: 621 pp., 2: 518 pp.

MIDDLETON, W. E. K., 1961. Pierre Bouguer's optical treatise on the gradation of light (translation, introduction, and notes). University of Toronto Press, Toronto, 247 pp.

MITCHELL, A. C., 1932. Chapters in the history of terrestrial magnetism, I. On the directive property of a magnet in the earth's field and the origin of the nautical compass. Terrestrial Magnetism and Atmospheric Electricity, 37: 105-146.

MITCHELL, A. C., 1937. Chapters in the history of terrestrial magnetism. The discovery of magnetic declination. Terrestrial Magnetism and Atmospheric Electricity, II-42: 241-280.

MITCHELL, A. C., 1939. Chapters in the history of terrestrial magnetism. The discovery of magnetic inclination. Terrestrial Magnetism and Atmospheric Electricity, III-44: 77-80.

MITCHELL, A. C., 1946. Chapters in the history of terrestrial magnetism. Terrestrial Magnetism and Atmospheric Electricity. IV-51: 323-351.

NEEDHAM, J., 1959. Science and civilization in China. University Press, Cambridge, 3: 874 pp.

NEEDHAM, J., 1965a. Science and civilization in China. University Press, Cambridge, 4(1): 434 pp.

NEEDHAM, J., 1965b. Science and civilization in China. University Press, Cambridge, 4(2): 759 pp.

NEWTON, I., 1687. The mathematical principles of natural philosophy. Reprinted 1964. The Citadel Press, New York, 447 pp.

OLMSTED, J. W., 1942. The scientific expedition of Jean Richer to Cayenne (1672-1673). Isis, 34: 117-128.

PARTINGTON, J. R., 1964. A history of chemistry. MacMillan and Company, London, 1: 370; 2: 795; 3: 854; 4: 1007.

PISARZHEVSKY, O. N., 1954. Dmitry Ivanovich Mendeleyev, his life and work. Foreign Languages Publishing House, Moscow, 102 pp.

POLANYI, M., 1958. Personal knowledge. University of Chicago Press, Chicago, 428 pp.

POSTAN, M. M. and HABAKKUK, H. J. (Editors), 1966. Cambridge economic history of Europe. 6 volumes.

POUTHAS, C., 1964. The revolutions of 1848, pp. 389-415. In: J.P.T. BURY (Editor), The New Cambridge Modern History, The Zenith of European Power, University Press, Cambridge, X: 766.

PRATT, J. H., 1855. On the attraction of the Himalaya Mountains and of the elevated regions beyond upon the plumb-line in India. Transactions of the Royal Society of London, B, vol. 145: 53-100.

PRATT, J. H., 1860. Figure of the earth. Macmillan and Company, London, 126 pp.

PRICE, D. J. de Solla, 1957. Precision instruments to 1500. In: C. SINGER et al (Editors), A History of Technology, Clarendon Press, Oxford, 3: 765 pp.

PRICE, D. J. de Solla, 1959. On the origin of clockwork, perpetual motion devices and the compass. Contributions from the Museum of History and Technology. U.S. Government Printing Office, Washington, D.C., paper 6, pp. 82-112.

PRICE, D. J. de Solla, 1961. Science since Babylon. Yale University Press, New Haven, 149 pp.

PRICE, D. J. de Solla, 1965. Little science, big science. Columbia University Press, New York, 119 pp.

ROYAL SOCIETY OF LONDON, 1909. Index, 1800-1900. 1: Mathematics, 2: Mechanics, 3(1): Physics (general heat, light, sound), 3(2): Physics (electricity and magnetism.

SCHMIDT, O., 1958. A theory of earth's origin. Foreign Languages Publishing House, Moscow, 139 pp.

SCHNEER, C., 1954. The rise of historical geology. Isis, 45: 256-267.

SCHNEER, C., 1967. Toward a history of geology. MIT Press, Cambridge, Massachusetts, 469 pp.

SEEGER, R. J., 1966. Galileo Galilei, his life and works. Pergamon Press, Oxford, 286 pp.

SINGER, C., 1957. Cartography, survey and navigation to 1400. In: SINGER et al.(Editors), A history of Technology, 3: 501-557.

SKELTON, R. A., 1958. Cartography. In: C. SINGER et al. (Editors), A History of Technology, Clarendon Press, Oxford, 5: 596-628.

SMITH, P. J., 1968. Pre-Gilbertian Conceptions of terrestrial magnetism. Tectonophysics, 6: 499-510.

SMITH, P. J. and NEEDHAM, J., 1967. Magnetic declination in mediaeval China. Nature, 214: 1213-14.

SOCIÉTÉ HOLLANDAISE, 1932a. Oeuvres completes de Christiaan Huygens, Martinus Nijhoff, The Hague, 17: 550 pp.

SOCIÉTÉ HOLLANDAISE, 1932b. Oeuvres completes de Christiaan Huygens, Martinus Nijhoff, The Hague, 18: 702 pp.

TAYLOR, E. G. R., 1937. The geographical ideas of Robert Hooke. Geographical Journal, 89: 525-538.

TAYLOR, E. G. R., 1948. The English worldmakers of the seventeenth century and their influence on the earth sciences. The Geographical Review, 38: 104-112.

THOMAS, R. H., 1951. Liberalism, nationalism and the German intellectuals (1822-1847). Heffer and Sons, Cambridge, 148 pp.

THOMSON, W. (Lord Kelvin), 1863. On the rigidity of the earth, shiftings of the earth's instantaneous axis of rotation; the irregularities of the earth as a timekeeper. Philosophical Transactions of the Royal Society of London. (Reprinted in: Mathematical and Physical Papers 1890, C. J. Clay & Sons, Cambridge University Press, London, 3: 312-350.

THOMSON, D., 1964. The United Kingdom and its worldwide interests, pp. 331-356. In: J.P.T. BURY (Editor) The New Cambridge Modern History, The Zenith of European Power, University Press, Cambridge, X: 766.

TODHUNTER, I., 1873. A history of the mathematical theories
of attraction and the figure of the earth. MacMillan
and Company, London. Reprinted 1962, Dover Publications,
New York, 1: 1-476, 2: 1-508.

TODHUNTER, I., 1886. A history of the theory of elasticity
and of the strength of materials from Galilei to
Lord Kelvin. University Press, Cambridge, 1: 936 pp.
(edited by Karl Pearson). Reprinted 1960, Dover
Publications, New York.

TODHUNTER, I., 1893. A history of the theory of elasticity
and of the strength of materials from Galilei to
Lord Kelvin. University Press, Cambridge. 2(1): 762;
2(2): 546.
Reprinted 1960, Dover Publications, New York.

TOMKEIEFF, S. I., 1950. James Hutton and the philosophy
of geology. Proceedings of the Royal Society of
London, Edinburgh, 63B: 387-400.

TURNBULL, H. W., (Editor), 1960. The correspondence of
Isaac Newton. The University Press, Cambridge, 2: 552 pp.

VAN BEMMELEN, R. W., 1967. The importance of the geonomic
dimensions for geodynamic concepts. Earth Science
Reviews, 3: 79-110.

VON HUMBOLDT, A., 1858. Cosmos. Henry G. Bohn, London, 5: 550 pp

VON ZITTEL, K. A., 1901. History of geology and
palaeontology. Walter Scott Press, London, 562 pp.

VYVYAN, J. M. K., 1960. Russia in Europe and Asia. In:
J.P.T. BURY (Editor), The New Cambridge Modern History,
The Zenith of European Power 1830-1870. University
Press, Cambridge, X: 357-388.

WEEKS, M.E., 1968. Discovery of the elements (7th edition).
Journal of Chemical Education, Easton, Pa., 898 pp.

WILSON, J. T., 1966. A theory of planetary behaviour.
Indian Geophysical Union, Hyderabad

NAME INDEX

SUBJECT INDEX

294